Alfred C(Alfred Charles) Garratt

Myths in Medicine and Old-Time Doctors

Alfred C(Alfred Charles) Garratt

Myths in Medicine and Old-Time Doctors

ISBN/EAN: 9783337184179

Printed in Europe, USA, Canada, Australia, Japan

Cover: Foto ©berggeist007 / pixelio.de

More available books at **www.hansebooks.com**

MYTHS IN MEDICINE

AND

OLD-TIME DOCTORS

BY

ALFRED C. GARRATT, M.D.

FELLOW OF THE MASSACHUSETTS MEDICAL SOCIETY; MEMBER OF THE
AMERICAN MEDICAL ASSOCIATION; MEMBER OF THE BOSTON
MEDICAL SOCIETY

" He comes to you with a tale which holdeth children from play, and old men from
the chimney corner."—SIR PHILIP SIDNEY

NEW YORK AND LONDON

G. P. PUTNAM'S SONS

The Knickerbocker Press

1884

Press of
G. P. Putnam's Sons
New York

CONTENTS.

INTRODUCTION.

THESE chapters record medical history, and the successive medical schools and sects; also the medical improvements along the ages, excepting the present,—these being the gleanings of special studies in medical history. The work is addressed to the profession, especially the younger portion of it, and to an intelligent public, in plain language for all readers. Many a beautiful and correcting thought, many a wise and helpful suggestion come down to us through these quaint old stories. They help us to eliminate myth and fallacy from the true, and show us scientific knowledge.

Here are traced the rise and progress of the great medical profession, the old and new healing art; also the new phase of a very old-fashioned medical " school," each inspected anew, by routes not usually taken, so that important facts that should be familiar to every one may be readily understood by any and all.

It is interesting to study the rise and fall of empires, and to discover their cause; so is it to observe the rise and struggling progress of art, science, philosophy, or mechanics, or any department thereof; to learn the cause of bias, or the hindering influences,—as from ignorance, or some pre-occupying false idea, or the fatal misleading of some foolish error introduced, it being first taught and elaborated perhaps by a mere visionary or conceited and erratic teacher, who became eminent only through the errors he led others to accept. The right and the wrong teachings of Aristotle and of Galen influenced the medical and the popular mind, their ideas and doings, for many ages. In like manner, and worse, the mere theoretical and chimerical teachings of Van Helmont, Paracelsus, and others have led astray a faction of the medical world, besides multitudes of other people, for ages more.

Having been much in the habit for over forty years, from the necessity of his specialty, of examining recent medical literature, especially that which aims to ascertain the cause, and point out the cure, of nervous diseases, the author was now and then led to search also well back into the earlier experiences of physicians in other ages, and in other states of society; which has impressed him with the curiosity and interest that would be elicited in this subject by the public generally if they could but see it. Why should not otherwise intelligent ladies and gentlemen have access to this line of ancient literature, and to the story of "medical improvement" through the ages? It is an accomplishment to know how to distinguish between reasonable and unreasonable expectations from the physician or his remedies; to discriminate between the true and false physician, as well as between true and false remedies; this concerns everybody.

The preparing, for popular reading, of this series of sketches of "The Old-school Doctors," and of medical writers of antiquity, and especially the showing how long continued was the want of a correct knowledge of anatomy and of physiology,—of *real nerves*, their nature and office, —and of various affections, their cause and cure, is but answering the oft-repeated queries from the laity on this subject that every practitioner has had to hear. The more correctly and fully informed the individual or community is in real medical matters the greater is the honest and hearty respect for the true physician and his prescription. And this confidence tends to save health and life. The last chapter of this little book logically grew out of the discussion of the real *old-school doctor* and medical dogmatism, from the earliest times to the latest.

To be ignorant is lamentable, but to be misled is dangerous and degrading. It is not honorable, nor just to the present or rising generation, for physicians and other cultivated and respectable bodies, or for individuals, now in the midst of the electric blaze of modern science, humanity, and Christianity, to condone, countenance, or even tolerate by silent connivance, a subtle and dangerous *organized error*, that presumes to claim and assume the attitude of being a part of the great medical profession; in which fact, every individual in society is interested. All true physicians are now called upon to improve every opportunity to instruct the people, and so correct popular medical fallacies.

In this book, then, three principal subjects are presented for the consideration of the reader. First, under what circumstances were nervous affections first described, and what were supposed to be their cause and nature, and what was the treatment in those times? Second, when did the general revival of *rational* medical research commence, which resulted in the present radical reformation, excellence, and unity of the regular profession into district societies, and these into state and national associations, and these into an international congress? Third, whether the often-repeated epithet "old-school" applies now to the regular medical profession of the present time, or rather to the strictly "dogmatic schools" and sects of the past and present?

Finally, this may show us from whence, how far, and how fast we have come on in the healing art ; and it may also serve as a background to the very numerous and beautiful word-pictures to be seen in various medical and other literature of the day, showing the marvellous and multitudinous " medical improvements," as well as preventive medicine ; all of which every person needs to know about, and many now rejoice in.

THE DOCTOR'S TALE.

"' YEA, let that passe,' quoth our Host, ' as now,
Sir Doctor Physic, I prayë you,
Tell us a Tale of some honest mattére.'
' It shall be done, *if that ye will it hear,*'
Saith this Doctor ; and his tale gan anon.
' Now good men,' quoth he, ' hearken every one.

' But first I pray you of your courtesy,
That ye arette it not my villainy,
Though that I plainly speak in this mattére.
.　.　.　.　.　.　.　.
For this ye knowen all, so well as I,
Whoso shall tell a tale after a man,
He must rehearse, as nigh as ever he can,
Every word, if it be in his charge,
Or elles he must tell his tale untrue.
He may not spare, although he were his brother ;
.　.　. He must as well say one word as another.
Eke Plato saith, whoso that him can read,
The wordie must be cousin to the deed.
Also I pray you to forgive it me,
All have I not set folk in their degree,
Here in this tale, as that they shoulden stand ;
My wit is short, ye may well understand."

Prologue in Canterbury Tale.

" OF all people who ever lived, the Persians were perhaps the most remarkable for their credulity and unshaken faith in amulets, spells, periapts, and similar charms, framed, it was said, under the influence of particular planets, and bestowing high medical powers, as well as the means of advancing men's fortunes in various manners.* A story, relating to a Crusader of eminence, is often told in the west of Scotland, and the relic alluded to, the Talisman, is still in existence, and is even yet held in veneration.

"Sir Simon Lockhart of Lee and Cortland made a considerable figure in the reign of Robert the Bruce and of his son David. He was one of the chief of that band of Scottish chivalry who accompanied James, the good Lord Douglas, on his expedition to the Holy Land. Douglas, impatient to get at the Saracens, entered into war with those of Spain, and was killed there. Lockhart proceeded to the Holy Land with such Scottish knights as had escaped the fate of their leader, and assisted for some time in the wars against the Saracens. The following adventure is said by tradition to have befallen him :—

" He made prisoner in battle an emir of considerable wealth and consequence. The aged mother of the captive came to the Christian camp, to redeem her son from his state of captivity. Lockhart is said to have fixed the price of ransom ; and the lady, pulling out a large embroidered purse, proceeded to tell down the ransom, like a mother who pays little respect to gold in comparison of her son's liberty. In this operation, a pebble inserted in a coin, some say of the Lower Empire, fell out of the purse, and the Saracene matron testified so much haste to recover it, as gave the Scottish knight a high idea of its value, when compared with gold or silver. ' I will not consent,' he said, ' to grant your son's liberty, unless that amulet be added to his ransom.' The lady not only

* " The Talisman," by Sir Walter Scott, p. 7.

consented to this, but explained to Sir Simon Lockhart the mode in which the Talisman was to be used, and the uses to which it might be put. The water in which it was dipped operated as a styptic, as a febrifuge, and possessed several other properties as a medical talisman.

"Sir Simon Lockhart, after much experience of the wonders which it wrought, brought it to his own country, and left it to his heirs, by whom, and by Clydesdale in general, it was, and is still, distinguished by the name of the 'Lee-penny,' from the name of his native seat of Lee.

"The most remarkable part of its history, perhaps, was, that it so especially escaped condemnation when the Church of Scotland chose to impeach many other cures, which savored of the miraculous, as occasioned by sorcery, and censured the appeal to them, 'excepting only that to the amulet called the Lee-penny, to which it had pleased God to annex certain healing virtues, which the Church did not· presume to condemn.' It still, as has been said, exists, and its powers are sometimes resorted to. Of late, they have been chiefly restricted to the cure of persons bitten by mad dogs ; and as the illness in such cases frequently arises from imagination, there can be no doubt that water which has been poured on the Lee-penny furnishes a congenial cure."

Such is the tradition concerning the Talisman that came from the Saracens, where it played a prominent part in their magical healing of the sick. The following example of its use is given :—

"'My lord,' said Kenneth, 'in plain language, then, I bring with me a Moorish physician, who undertakes to work a cure on King Richard.'

"'A Moorish physician !' said De Vaux ; 'and who will warrant that he brings no poison instead of remedies ?'

"'His own life, my lord—his head, which he offers as a guarantee. Thus it is, my lord, that Saladin, to whom none will deny the credit of a generous and valiant enemy, has sent his leech hither with an honorable retinue and guard, befitting the high estimation in which El Hakim is held by the Soldan, and with fruits and refreshments for the King's private chamber, and such message as may pass betwixt honorable enemies, praying him to be recovered of his fever, that he may be the fitter to receive a visit from the Soldan, with his naked scimitar in his hand, and an hundred thousand cavaliers at his back. Will it please you, who are of the King's secret council, to cause these camels to be discharged of their burdens, and some order taken as to the reception of the learned physician?'

"'Wonderful!' said De Vaux, as speaking to himself. 'And who will vouch for the honor of Saladin, in a case when bad faith would rid him at once of his most powerful adversary?'

"'I myself,' replied Sir Kenneth, 'will be his guarantee, with honor, life, and fortune.' . . . .

"'I have been absent on a pilgrimage, in the course of which,' replied Sir Kenneth, 'I had a message to discharge towards the holy hermit of Engaddi.'

"'Tell me, Sir Knight of the Leopard, granting (which I do not doubt) that thou art thyself satisfied in this matter, shall I do well, in a land where the art of poisoning is as general as that of cooking, to bring this unknown physician to practise with his drugs on a health so valuable to Christendom?'

"'My lord,' replied the Scot, 'thus only can I reply: that my squire and attendant has been of late suffering dangerously under the same fever which, in valiant

King Richard, has disabled the principal limb of our holy
enterprise. This leech, this El Hakim, hath ministered
remedies to him not two hours ago, and already he hath
fallen into a refreshing sleep. That he *can* cure the disorder, which has proved so fatal, I nothing doubt ; that
he hath the purpose to do it is, I think, warranted by
his mission from the royal Soldan, who is true hearted
and loyal, so far as a blinded infidel may be called so ;
and, for his eventual success, the certainty of reward in
case of succeeding, and punishment in case of voluntary
failure, may be a sufficient guarantee.'

" The Englishman listened with downcast looks, as
one who doubted, yet was not unwilling to receive conviction. At length he looked up and said, ' May I see
your sick squire, fair sir ?' . . . .

" Beside the couch sat, on a cushion, also composed of
skins, the Moorish physician, of whom Sir Kenneth had
spoken, cross-legged, after the Eastern fashion. . . .
The English lord entered and stood silent, with a sort of
reverential awe. Nothing was for a time heard but the
heavy and regular breathings of the invalid, who seemed
in profound repose.

"'He hath not slept for six nights before,' said Sir
Kenneth, 'as I am assured by his attendants.' He had
scarcely uttered these words, when the physician, arising
from the place which he had taken near the couch of the
sick, and laying the hand of the patient, whose pulse he
had been carefully watching, quietly upon the couch,
came to the two knights, and taking them each by the
arm, while he intimated to them to remain silent, led
them to the front of the hut.

" He said : ' Disturb not the effect of the blessed medicine of which he hath partaken. To awaken him now, is
death or deprivation of reason ; but return at the hour

when the Muezzin calls from the minaret to evening prayer in the mosque, and, if left undisturbed until then, I promise you, this same Frankish soldier shall be able, without prejudice to his health, to hold some brief converse with you.' . . . .

"'This is a strange tale,' said the sick monarch, when he had heard the report of the trusty men. . . . .

"'And did they meet the physician?' demanded the king, impatiently. . . . .

"'No, my liege; but the Saracen, learning your Majesty's grievous illness, undertook that Saladin should send his own physician to you, and with many assurances of his eminent skill. He has come, and is attended as if he were a prince, with drums and atabals, and servants on horse and foot, and brings with him letters of credence from Saladin.' . . . .

" Richard took a scroll, in which were inscribed these words:—

"'The blessing of Allah and his Prophet Mohammed; Saladin, king of kings, Soldan of Egypt and of Syria, the light and refuge of the earth, to the great Melech Ric, Richard of England, greeting: Whereas we have been informed that the hand of sickness hath been heavy upon thee, our royal brother, and that thou hast with thee only such Nazarine and Jewish mediciners as work without the blessing of Allah and our holy Prophet, we have therefore sent to tend and wait upon thee at this time the physician to our own person, Adonbec el Hakim, before whose face the angel Azrael (Death) spreads his wings, and departs from the sick chamber; who knows the virtues of herbs and stones, the paths of the sun, moon, and stars ; and can save man from all that is not written on his forehead. And this we do, praying you heartily to honor and make use of his skill.'

. . . . "'Hold, hold,' said Richard, 'I will have no more of this. It makes me sick to think the valiant Soldan should believe in a dead dog.—Yes, I will see his physician. I will put myself into the charge of this Hakim. I will repay the noble Soldan his generosity—I will meet Saladin in the field, as he proposes, and he shall have no cause to term Richard of England ungrateful. I will strike him to the earth with my battle-axe— I will convert him to Holy Church with such blows as he has rarely endured. He shall recant his errors before my good cross-handed sword, and I will have him baptized in the battle-field, from my own helmet, though the cleansing waters were mixed with the blood of us both. Fetch the Hakim hither.' . . . .

"The baron De Vaux and the Archbishop of Tyre said to the physician: 'We would have oracular proof of thy skill, and without it thou approachest not to the couch of King Richard.'

"'The praise of the physician,' said the Arabian, 'is in the recovery of his patient. Behold this sergeant, whose blood has been dried up by the fever which has whitened your camp with skeletons, and against which the art of your Nazarine leeches hath been like a silken doublet against a lance of steel. Look at his fingers and arms, wasted like the claws and shanks of the crane! Death had this morning his clutch on him; but had Azrael been on one side of the couch, I being on the other, his soul should not have been reft from his body. Disturb me not with further questions, but await the critical minute, and behold in silent wonder the marvellous event.'

"The physician had then recourse to his *astrolabe*, the oracle of Eastern science, and, watching with grave precision until the precise time of the evening prayer had arrived, he sunk on his knees, with his face turned to

Mecca, and recited the petitions which close the Mosleman's day of toil. The bishop and the English baron looked on each other, meanwhile, with symptoms of contempt and indignation, without interrupting El Hakim in his devotions.

"The Arab arose from the earth, on which he had prostrated himself, and, walking into the hut where the patient lay extended, he drew a sponge from a small silver box, dipped perhaps in some aromatic distillation ; for when he put it to the sleeper's nose, he sneezed, awoke, and looked wildly around. He was a ghastly spectacle, as he sat up almost naked on his couch, the bones and cartilages as visible through the surface of his skin, as if they had never been clothed with flesh; his face was long, and furrowed with wrinkles. He seemed to be aware of his dignified visitors, for he attempted feebly to pull the covering from his head, in token of reverence. . . .

"'Your eyes witness,' said the Arabian, 'the fever has been subdued—he speaks with calmness and recollection —his pulse beats quietly.'

"'This is most wonderful,' said the knight, looking to the bishop ; 'the sick man is assuredly cured. I must conduct this mediciner presently to King Richard's tent, What thinks your Reverence ?'

"'Stay, let me finish one cure ere I commence another,' said the Arab ; 'I will pass with you when I have given my patient the second cup of this most holy elixir.'

"So saying, he pulled out a silver cup, and filled it with water from a gourd which stood by the bedside. He next drew forth a small silken bag made of network, twisted with silver, the contents of which the bystanders could not discover, and immersing it in the cup, continued to watch it in silence, during the space of five minutes. It seemed to the spectators as if some effervescence took place during the operation, but if so, it instantly subsided.

" 'Drink,' said the physician to the sick man—'sleep, and awaken free from malady.'

" 'And with this simple-seeming remedy thou wilt undertake to cure a monarch?' said the Bishop of Tyre.

" 'I have cured a beggar, as you behold,' replied the sage. 'Are the Kings of Frangistan made of other clay than the meanest of their subjects?'

" 'Let us have him presently to the King,' said the baron. 'He hath shown that he possesses the secret, which may restore his health. If he fails to exercise it, I will put himself past the power of medicine.' . . .

" 'This is the Prince of Leeches ; fever, plague,
Cold rheum, and hot podagra, do but look on him,
And quit their grasp upon the tortured sinews.'

"Richard, when they entered his apartment, immediately exclaimed, 'So ho! a goodly fellowship come to see Richard take his leap in the dark. My noble allies, I greet you as the representatives of our assembled league ; Richard will again be amongst you in his former fashion, or ye shall bear to the grave what is left of him. —Come, Sir Haikim, to the work, to the work.'

"The physician, who had already informed himself of the various symptoms of the King's illness, now felt his pulse for a long time, and with a deep attention, while all around stood silent, and in breathless expectation. The sage next filled a cup with spring water, and dipped into it the small red purse, which, as formerly, he took from his bosom. When he seemed to think it sufficiently medicated, he was about to offer it to the sovereign, who prevented him, by saying, " Hold an instant. Thou hast felt my pulse—let me lay my finger on thine; I too, as becomes a good knight, know something of thine art.'

"The Arab yielded his hand without hesitation. ' His blood beats calm as an infant's,' said the King; ' so throb not theirs who poison princes.' . . . .

"He then raised himself in bed, took the cup in his hand, and said: ' To the immortal honor of the first Crusader who shall strike lance or sword on the gate of Jerusalem.' He drained the cup to the bottom, then resigned it to the Arabian, who then, with silent but expressive signs, directed that all should leave the tent; and the apartment was soon cleared.

"The critical hour arrived at which the physician, according to the rules of his art, had predicted that his royal patient might be awakened with safety, and the sponge had been filled and applied for that purpose; and the leech had not made many observations ere he assured the baron that the fever had entirely left his sovereign, and that such was the happy strength of his constitution, that it would not be necessary, as in most cases, to give a second dose of the powerful medicine. Richard himself seemed to be of the same opinion, for, sitting up and rubbing his eyes, he demanded of De Vaux what present sum of money was in the royal coffers.

"'It matters not,' said Richard; 'be it greater or smaller, bestow it all on this learned leech, who hath, I trust, given me back again to the service of the Crusade. If it be less than a thousand byzants, let him have jewels to make it up.'

"'I sell not the wisdom with which Allah has endowed me,' answered the Arabian physician, 'and be it known to you, great Prince, that the divine medicine, of which you have partaken, would loose its effects in

2

my unworthy hands, did I exchange its virtues either for gold or diamonds.' . . . .

" ' Explain thy words,' said Richard. El Hakim answered, ' Know thou that the medicine to which thou, Sir King, and many one beside, owe their recovery, is a *talisman*, composed under certain aspects of the heavens, when the Divine Intelligences are most propitious. I am but the poor administrator of its virtues. I dip it in a cup of water, observe the fitting hour to administer it to the patient, and the *potency* of the draught works the cure.'

" ' A most rare medicine,' said the King, ' and a commodious! and as it may be carried in the leech's pocket, or purse, would save the whole caravan of camels which they require to convey drugs and physic-stuff. I marvel there is any other in use.'

" ' Know, that such talisman might indeed be framed,' said Hakim, ' but rare has been the number of adepts who have dared to undertake the application of their virtues. . . . When thou canst show why that draught of cold water should have cured thee, when the most precious drugs failed, thou mayest reason on the other mysteries attendant on this matter.' . . . .

" The next day saw Richard's return to his own camp, and in a short time afterwards, the young Earl of Huntingdon was espoused by Edith Plantagenet. The Soldan himself, who was the physician in disguise, sent as a nuptial present on this occasion the celebrated *Talisman ;* but though many cures were wrought by means of it, in Europe, none equalled in success or celebrity those which the Soldan achieved. It is still in existence, as before stated, having been bequeathed by the Earl of Huntingdon to a brave knight of Scotland, Sir Simon of the Lee, in whose ancient and highly honored family it is still preserved; and although charmed stones have been dismissed from the modern pharmacopœia, *its* virtues are still applied to, for stopping blood, and in cases of canine madness." *

* " The Talisman," by Sir Walter Scott, p. 287.

CHAPTER I.

EMINENT PHYSICIANS IN ANCIENT TIMES, FROM HIP-POCRATES TO GALEN.

At the present time the word improvement is as familiar to physicians as it is to mechanics. A book for guide in medical practice that has been published over a dozen years, is looked upon as old, and needs to be read with amendments. Those written fifty or one hundred years ago are considered not only old, but obsolete, while those that are two or three hundred years old are, in this country, rare to be found. Probably not one man in a hundred has ever seen, and certainly does not own, a copy of any such medical work. At the same time, it is true that a few very select works, written in the more ancient times, are better known and valued as curiosities, if nothing more. The most of these contain many grains of choice wheat as well as bushels of chaff, showing a crude old practice, based on erroneous teachings and ignorance intermingled with some scattered facts.

Then we find that, the further along the course of time we examine the prominent authors on medicine, the more we are convinced that the lingering shadows of the dark ages still befogged medical teachings in a very marked degree, even in its best ranks, up to the seventeenth century; that is, to within one or two hundred years of our own day.

It was not until about 1600 A.D. that the old traditional and dogmatic empiricism began to be openly questioned, then doubted, and finally rejected, only by a few of the most learned in medicine and philosophy. Then how partially and slowly it went on! Still humoral pathology held pre-eminent sway in all the learned medical world everywhere for more than two hundred years. In the last fifty years there has been a rapidly increasing succession of improvements in all departments of the healing art; whose libraries, covering the entire field, have been written and re-written, again and again, during this time.

Quite an authentic account of the healing art, such as it was, reaches back along the last third of the history of man, which is but a little over two thousand years. True, much mystical medical art was already developed in Egypt, Assyria, and Greece, even more than one thousand years before that. But when we speak of connected and reliable records of medicine and surgery, as they existed in early times, we find we are indebted first of all to the great Greek physician and surgeon, Hippocrates. He lived about 400 B.C. and was the most systematic and extensive writer in medicine up to that time. With him began the true history of medicine. Yet he, and those of his time, knew but little of the anatomy and physiology of the human body, and but little of the true outlines of diseases, or of the range and power of remedies, and probably nothing of the true character of the nervous system.

Hippocrates prudently relied very much on what he called *nature,* and is known to us as the *vital force,* or vitality; also termed the *vis medicatrix natura.* Hence he dwelt much on the method of fasting and eating, of diet, regimen, hygiene, external remedies, and open air for exercise and sleep. Though he gave rules for medical study to this end, yet soon after he died medical ideas and practice relapsed again into a mixture of speculation, science and philosophy, astrology and alchemy, as of old, but in a less degree.   He long endeavored to separate the medical art from the priestly office, and from the oracles, and from the admixture of the prevailing philosophy of that time.   He was born nearly a century after Pythagoras, and inherited medicine as a profession.

Pythagoras, before assuming the business of teaching philosophy and medicine, had spent much time in travelling through many countries, and probably taught something of the Egyptian medical art in his courses of instruction at Carona, where medicine was first cultivated as a department of philosophy in Europe.   This school of Pythagoras at Carona in Magna Græcia, now the south part of Italy, preceded that of Hippocrates and that of Plato by more than a century.

" Now, to the initiated, medicine is something more than a profession, it is a world within itself.*   It has its history, its philosophy, its politics, its literature, of which the world at large knows nothing.   It has its subsidiary arts and occupations.   It has its organizations and institutions, its ranks and grades of honor.   It has its polemics and dissensions, not always amenable to logic or the learning of the academies.   In ethics, traditions, and superstitions it is older than the church.   In use before the civil law, it recognizes no arbitrary enactments. Nature is its only court of equity.   And who of us shall forget its ever-living charities, its many sunny aspects, its benignant, ennobling, liberalizing influences, which few beyond our own circle can properly appreciate, and none so well understand.

" No wonder, then, that the members of our profession, drawn together by these hallowed ties, should be disposed

------

* " History of Medicine," by Winslow, p. 10.

.ÆSCULAPIUS.

to band together as a brotherhood. Such has been their
tendency. The Druids of early Gaul and Britain, the
Asclepiadæ of Greece, the priests of ancient Egypt, the
Lamas of central Asia, the Vaidhyas of India, the frater-
nities of the middle ages, and up to the present hour the
countless societies and colleges of our own and other
lands, and associations, national and international, devoted
to the healing art, are in proof of this. So that where-
ever social freedom has existed, or tyranny would per-
mit, internal organization and development has been
the rule of our profession. With these facts before us,
our origin and growth as an element of civilization is a
subject worthy of general attention."

In the very early times, " there is reason to believe that
among the Assyrians and other early Asiatics medicine
was never pursued as a distinct occupation. The eastern
Magi must have devoted some attention to it; and the
seers of Palestine may have had some pretensions to skill
in curing diseases as a part of their divine calling. Job
speaks of his counsellors as ' physicians of no value,' and
Moses, of the preparation of the sacred oil after 'the
apothecary's art.' King Asa, when his disease was exceeding great, ' sought not the Lord, but his physicians ;'
and Jeremiah asks, ' Is there no balm in Gilead? is there
no physician there?' From these and other allusions in
the Old Testament, it is evident that among the Isra-
elites there were, perhaps after the manner of the Egyptians, certain men giving their attention to medicine.
But the Babylonians, as we learn from Herodotus,* were
destitute of physicians," as were also the kings of Persia,
only as they obtained them from Greece or Egypt.

Æsculapius himself, the ancestor of Hippocrates, is
said to have had two sons, who distinguished themselves

* Book I., par. 197.

as physicians and surgeons at the siege of Troy, in 1184
B.C. It is recorded that medical knowledge was in those
days retained as a *secret* in his family, it being transmitted from father to son until the time of Hippocrates
many ages after. The more immediate descendants of
Æsculapius formed a priesthood, a long and powerful
line, known as the " Asclepiadæ," who practised medicine from the Temples during many ages, and are frequently referred to by eminent medical authors for more
than two thousand years after.

> . . . " A cave y-wrought by wonderous art,
> Deep, dark, uneasy, doleful, comfortless,
> In which sad Æsculapius, far apart
> Imprisoned was in chains remediless,
> For that Hippolytus' rent corse he did redress,
>
> . . . His goodly corse, on ragged cliffts y-rent,
> Was quite dismembered, and his members chaste,
> Scattered on every mountain as he went.
>
> By Dyan's means who was Hippolyte's friend,
> Them brought to Æsculapius that by his art
> *Did heal them all again and joined every part.*
> Such wondrous science in man's wits to reign
> When Jove advised, that could the dead revive
> And fate expirèd could renew again,
> Of endless life he might not him deprive ,
> But unto hell did thrust him down alive,
> With flashing thunderbolt y-wounded sore :
> Where, long remaining, he did always strive
> Himself with salves, to health for to restore,
> And slake the heavenly fire that ragèd evermore.
>
> Beseeching him with prayers, and with praise,
> If either salve, or oils, or herbs, or charms,
> A foredone wight from door of death might raise,
> He would at her request, prolong her nephew's days.

. . . .

Go to, then, O thou far renowned son
Of great Apollo! show thy famous might
In medicine, that else hath to thee won
Great pains, and greater praise, *both never to be done.*"
                    Spenser's " Faery Queen."

Was he who was known as Æsculapius a Greek or an
Egyptian? It is believed he was a real character, who
in after times was deified by the Greeks, and represented
as the son of Apollo and Cornis. Greece was early colo-
nized by the Egyptians; then in after ages the Greeks
were led to colonize Egypt, carrying their arts and sciences with them, which so embellished the great and famous school of learning in Alexandria.

Hippocrates left many books of his own writing on
medicine and surgery, of which certainly five, and probably sixteen, if not many more, are genuine. We introduce here this long gone by and well known fact, because
from these books re-appear again and again certain leading, and many other very misleading doctrines, sometimes
modified this way or that, as appear in the best medical
works made along the ages, and show prominently in
books written by the learned in England and Europe,
even long after the Pilgrim Fathers landed on Cape Cod
in Massachusetts Bay. For this reason, too, are introduced in short the successive authors, schools, and sects
of very different times, to show the bewildering state
of medicine as a profession up to the eighteenth century.

Explorations recently made on the site of the Temple
of Æsculapius at Epidaurus in Argolis, have brought to
light five statues and many inscriptions, besides portions
of a celebrated stile, mentioned by Pausanias as standing near the Temple, and bearing the names of persons

who had been cured, with an account of their maladies, the remedies applied, the sacrifices offered, etc.

Let it be repeated, then, that what ideas Hippocrates had of the bones and organs of the human body were only general and comparative, and often quite erroneous.

Hippocrates, "The Prince of Medicine," "The Father of Physic," "The Oracle of Cos," "The Divine Old Man," was born at Cos, one of the islands of the Grecian Archipelago, at the commencement of the eightieth Olympiad, four hundred and sixty years B.C. His father was a physician, the seventeenth lineal descendant from Æsculapius. His mother was Phenerata, the eighteenth in descent from Hercules. In his family there were no less than seven physicians named Hippocrates. The first was the contemporary of Miltiades. Hippocrates II., the Great, is the father of medicine referred to. Hippocrates III. was grandson of the Great, and wrote works on medicine. Hippocrates IV. was physician at the court of Macedonia, and is said to have cured Roxana, widow of Alexander the Great.

Hippocrates lived in the best and most prosperous period of Greece. He was the contemporary of Socrates, Plato, and Herodotus, and he received his earliest knowledge of medicine from his father and from Herodicus. He also studied under Gorgias, the celebrated orator. Hippocrates wrote in the Ionic dialect, and he used many Attic expressions. There were great stores of medical records accumulated at the celebrated Temple of Æsculapius in the island of Cos, to which he had access. During his travels he also visited and sojourned at Ephesus, near the temple of Diana, where he transcribed and arranged the Tables of Medicine therein preserved; for it was the custom there to record every case, and the treatment, whatever was the result.

Hippocrates, we observed, founded his views of medicine on facts, or what he supposed were facts, observed or found in records of medicine. At his standpoint he learned and accepted from the schools of philosophy the doctrine of the primitive elements, and that of the primitive humors, and sees in the human body the humors undergoing changes in accordance with the conditions of health or disease. He was led to believe that health is maintained by the equable proportion and admixture of the humors, and that disease is the result of their inequalities;* that during their changes, when they happen to be unusually great, the disordered humors undergo a process of *coction*, by which they may be restored to their healthy condition; and as time is necessary to effect this process, he endeavors to show how the critical discharge is brought about, and to establish the very days within which it is to be expected. He lays much stress on the doctrine of coction, implying by this term the changes disordered humors undergo preparatory to their elimination. So long as they float about in a state of crudity the disease continues in full intensity; but it ceases when they are properly elaborated, either by the spontaneous effort of nature, or by the aid of medicine acting under natural laws.

Though he was evidently not aware of the real nerves or the nervous system, yet he mentioned the term "*nerve*," as Greek authors of those times usually did, to signify a sinew, or tendon. He distinguished between arteries and veins, but he supposed the brain to be a gland which exuded "a viscid fluid;" that the heart was "the fountain of life," having two "ventricles," or cavities, separated by a partition; and as also having two "auricles," as receptacles of air. He taught the

* "Medical Heresies," by Prof. G. C. Smythe, p. 33.

doctrine of *solids*, *humors*, and *spirits;* also that of the three humors of the blood—phlegm, yellow bile, and black bile. He also divided diseases into the acute and chronic, epidemic and sporadic, malignant and benign. His theory of diseases was based on the four elements,— earth, fire, air, and water,—together with the three humors of the blood. He believed that, as before stated, some kind of derangement of these elements and humors caused disease; and that through the indwelling nature, that nature or vitality to which he attributed real intelligence in distributing the blood, spirits, and warmth throughout the body and limbs, the body receives life and growth, sensation, motion, and nourishment during health, while in disease it produces " coction," or a ripening of the matters of disease, which are finally expelled at the " crisis."

The Oath of Hippocrates.

"I swear by Apollo the physician, by Æsculapius, by his daughters Hygeia and Panacea, and by all the gods and goddesses, that to the best of my power and judgment I will faithfully observe this oath and obligation. The master that has instructed me in the art I will esteem as my parent, and supply, as occasion may require, with the comforts or necessaries of life. His children I will regard as my own brothers ; and if they desire to learn, I will instruct them in the same art without any reward or obligation. The precepts, the explanations, or whatever else belongs to the art, I will communicate to my own children, to the children of my master, to such other pupils as have subscribed to the physician's oath, and to no other persons. My patients shall be treated by me to the best of my power and judgment, in the most salutary manner, without any injury or violence; I will neither be prevailed upon by any other to administer pernicious physic, or to be the author of such advice myself. Cutting for the stone I will not meddle with, but leave it to the operators in that way. To whatsoever house I am sent for, I will always make the patient's good my principal aim, avoiding, as much as possible, all voluntary injury and corruption. And, whatever I hear or see in the course of a cure, or otherwise, relating to the affairs of life, nobody shall ever know it, if it ought to remain a secret. May I be prosperous in life and business, and forever honored and esteemed by all men as I observe this solemn oath; and may the reverse of all this be my portion if I violate it, and forswear myself."—*Hippocrates.*

Soon after appeared Polybius, the son-in-law of Hippocrates, who in his later years wrote two medical works, interspersed with anatomy and physiology. To show the crude ideas in regard to the structure of the human body they had up to that time, I quote him: "There are four pair of large or main blood-vessels in the human body. The *first* pair lead from the parietal portions of the head down the sides of the neck and body to the outer parts of the thighs and ankles; the *second* pair, being the jugular veins, leading from the sides of the neck to the inner side of the thighs, and so on down to the inner side of the ankles and feet. The *third* pair lead from the temples of the head through the lungs, and there crossing each other, the one goes to the *spleen* and the left kidney, while the other goes to the liver and the right kidney. The *fourth* pair of blood-vessels lead from the throat and neck over the upper extremities, and over all the front part of the body." Such were evidently the anatomical views of the most advanced and eminent medical men in learned Greece to within three hundred and fifty years before the Christian era. It is a wonder their theory and practice was so often so nearly correct.

Next appeared most prominently the great Aristotle, who was a Greek, born in 384 B.C. He was a pupil of the greater Plato, and was afterwards his great expounder. Plato had been the pupil of the still greater Socrates.* Aristotle had a vast collection of facts, upon which he based much sound doctrine, and very much more purely chimerical nonsense, out of which grew his famous philosophy, the most plausible, comprehensive, and absurd system ever devised by man, which largely influenced medicine, religion, and the world of letters generally for more than two thousand years afterwards. By his actual dissections of animals he was the real founder of comparative anatomy and zoology.

By Aristotle was corrected some of the errors recorded by Polybius. Through him and such sages in medicine and philosophy as Dracho, Theophrastus, and Praxagoras, was the general stock of knowledge in anatomy and physiology enriched. Aristotle divided the circle of knowledge into four parts, namely, physics, metaphysics, logic (which included what we now term biology—the science of the mind), and ethics. He and others mingled his philosophy with the somewhat purified medical art of Hippocrates, and so no doubt retarded medical improvement to more than a thousand years after. Out of it grew the dogmatic sect, which was soon divided into three factions in the time of Galen, and they continued a long time.

It was known to the ancients that worms sometimes came out of the nose of animals, and they were believed

* Socrates, " Apology," " Cryto," and parts of the " Phædo " of Plato.

to cause the disease called " the staggers," or "turning-
fits," in sheep. Early medical history states that they
found these maggots, hatched from the eggs of a certain
fly, lodged in the frontal bone of the head. " No doubt
this is why the ancients believed that the larvæ from the
sheep's head were an effectual remedy in epilepsy. Of
course they were prescribed on the principle that what
produces a disease will cure that disease (according to
one of those very ancient medical aphorisms). As early
as B.C. 560 Alexander Trallianus tells us that at two distinct utterances the Oracle of Delphi recommended these
worms to be used by Democrates of Athens, who suffered
with epilepsy. But as Democrates knew nothing of natural history he asked a man one hundred years old, who
told him to take the worms which fell from the nose of
such sick sheep, and tie them up in a bag and hang them
about his neck."

Truly there is nothing new under the sun—"cure like
with like" in the year B.C. 650! " The priestcraft, preach-
ing through the Oracle of Delphi to the people, repeats
itself in our day in the Oracle or Oganon of Hahnemann,
the father of Homœopathy. Strange that in our enlightened time, when science in all branches of learning has
given us such practical and valuable information, that
there still remains in some of our highly educated, and,
in other respects, so practical citizens, *that mystical belief*
(that has filtered through the dark ages), as stated in the
garb of Homœopathy."—*F. Humbert, M.D.*

In B.C. 320, or about one hundred years after the time
of good Hippocrates, was founded the great school of
learning at Alexandria in Egypt, under the protection
and patronage of the Ptolemies ; and there and then
grew up the great " Empirical" school of medicine, the
rival of the Dogmatics, with which they divided the med-

ical world for centuries. This school of empirics rejected
the doctrine of "occult causes" of disease, as had been
taught by Hippocrates, and afterwards adopted by the
Dogmatics, while they based their system entirely on
observation and experience gained through the senses.
They also rejected the very old axiom, that diseases
must be cured by *contraries;* that is, by remedies that
produce in the human body opposite or different effects
from that produced by the given disease. As one of the
axioms in medicine, this no doubt is very ancient; it being the most simple and natural idea that would first occur to the mind of man, as indicating the sort of remedy
that should be selected for relieving and curing the given
sickness. But it did not make any exclusive *sect* then,
nor has it ever done so to the present time.

In later ages we find three or four of these axioms, or
medical aphorisms, in use as antitheses, employed mostly
in didactics, when speculating on the mode of action of
different kinds of remedies, namely, "antipathy," which
means a natural aversion, or opposition ; "allopathy,"
which indicates a dissimilarity, reversion, or contrary ;
and "homœopathy," which indicates alike, as this thing
and that are alike. What is strange to relate, some one
hundred years after these axioms had mostly disappeared
from medical literature, and different sects and schools
were dying out, Hahnemann separated himself from the
regular medical profession, and proclaimed, as a universal
and exclusive law, that "like cures like," and that all else
is false and wrong; then he adopted the creed, *similia
similibus curanter*, and called his sect Homœopathy.

About the year 50 B.C. was established the "Me-
thodic" school of medicine by Asclepiades, who was
born in Bithinia, and practised in Rome, being an inti-
mate friend of Cicero, and a favorite of the nobles of

that city. He taught that the human body was every-where filled with minute pores, and that invisible cor-puscles, floating in fluid, were constantly passing through these pores; that the food was first simply emulsified, then passed directly into the blood, where it was further fined, so as to pass easily through these filtering passages and so act as nutriment. He also taught that hunger was caused by the relaxation of the larger pores, thirst by the relaxation of the smaller pores; that sicknesses were caused by spasm of the pores, or else the corpus-cles, not being well concocted, remained too large to pass through the ordinary pores, or the pores themselves had become relaxed. Hence the aphorism of this school, in short, was *Relaxum et Strictum*, as afterwards the ho-mœopathic school had " similia similibus curanter." This was about the whole of it; and consequently the reme-dies were friction, wine, exercise, bathing, with shampoo-ing and injections.

Men, yes, eminent men, in those days believed in this; such as Themison of Laodicea, Stephenson of Bizantium, Celeus and Marcus Artorius of Rome, and Rufus of Ephe-sus, while the emperor, Augustus Cæsar himself, was his patron. In time Atheneus added to this *new creed* the other idea of spirits, or air passing through the arteries, which, when impeded, or otherwise disturbed in the body, caused disease; which doctrine then gave rise to the "Pneumatic" school of medicine, prominent in which was the distinguished physician Arteus. This sect after a time broke into several factions, each having but a partial following out of the medical profession as a whole.

Soon after the Christian era, the more prominent phy-sicians of all sects and factions came together in an agreement that no one faction contained all the truth,

3

therefore they would unite in forming the famous " Eclectic school " of medicine, which was expected then to embrace the medical world, for they claimed to select all that was good and useful, and reject all else. But it was too early for such a move. Still the various sects remained here and there numerous and divided for a very long time, until Galen's day, which was an epoch in medical improvement, unity, and history.

Erasistratus, professor in the great college of Alexandria, held medicine to be a conjectural science. He did not entirely reject the pharmaceutical branch of medicine, but he opposed venesection, purgation, and most other active medicines; he treated disease almost exclusively by diet, regimen, and hygiene. On the other hand, Herophilus, his brother professor at Alexandria, and his followers, extolled medicines, and resorted to them in all diseases. Some of these wrote extensively on the " materia medica." Among the earliest were Zeno, Andreas, and Appolonias. It was Pamphilius who wrote on herbs, giving a description of their medical use. Compiling from Hermes, the great medical authority of the early Egyptians, he dwells upon the use of charms, amulets, and incantations, " for increasing the power or potency of herbs." Nicander, who lived in those days at Pergamus, wrote a treatise on the bites and stings of venomous animals and poisons.

Attalus Philometor was not only the patron of the medical profession, but was himself actively occupied in the cultivation and administration of medicinal plants. We are told by Plutarch, that in his gardens he planted the cicuta, aconite, hellebore, hyoscyamus, and other active herbs.* " He was particular to collect them at certain seasons, with his own hands, for the purpose of

* Le Clerc, part ii., liv. iii., chap. iii., p. 388.

testing and experimenting with their expressed juice,
their fruits, and their seeds and roots, and determining
their respective properties." The king of Pontus, Mith-
ridates, also experimented with poisonous plants, upon
himself, and finally composed the celebrated " *mithridat-
icum,*" employed as an antidote, which was among the
most famous nostrums of antiquity. Pompey, after over-
coming this prince, ordered diligent search to be made
among the archives of his palace for the formula of this
famous antidote, and thought he found it.

" In speaking of the medical profession in his own time
Galen classes them largely as Herophilians and Erasis-
tratians, showing that the opinions of the founders of the
Alexandrian school in Egypt had not yet been super-
seded, and that after an interval of more than four centuries, the impression left upon it by these great men
still continued to give it character and distinction.* After
falling under the sway of the Romans for a century or
more the school of Alexandria lost much of its previous
celebrity, and is little spoken of during the more active
period of the medical school of Rome. Yet, even in
Galen's day, it was the centre of medical science. For
to have studied medicine at Alexandria was everywhere
considered a passport to the confidence and patronage of
the public in any country."

Erasistratus, who was a native of the Isle of Chios,
and the grandson of Aristotle, flourished about two hundred and fifty years before the Christian era. He was
one of the most prominent of the scientific men brought
together at the Museum in Alexandria by Ptolemy
Soter, its founder, who was by repute the natural son of
Philip of Macedon, and therefore brother of Alexander
the Great. The Museum had already risen to the high-

* Watson's " Medical Profession," p. 94.

est ranks among the Greek schools. The library at that time held two hundred thousand rolls of papyrus, equal to about ten thousand of our modern printed volumes. The system of instruction, as first arranged, was divided among the four faculties of literature, astronomy, mathematics, and medicine. At the head of the latter were Erasistratus, Cleombrotus of Cos, and Herophilus the anatomist. Cleombrotus was in high repute as a practitioner; was sent to the relief of Antiochus when dangerously ill, and after curing the king he received on his return a hundred talents, about fifteen thousand pounds sterling, as a reward, from Ptolemy Philadelphus.

Erasistratus, who paid no regard to the Hippocratic doctrine of humors or elements, made discoveries in anatomy and physiology, and wrote extensively. He was familiar with some of the general distributions of the blood-vessels. He described the anatomical structure of the heart, and, like Aristotle, made this organ the source both of the veins and arteries. He held also that the arteries in health are filled with pneuma, or air, received by the process of respiration, and that the passage of blood into them from the veins is the usual cause of disease. Rufus of Ephesus says Erasistratus was somewhat familiar with the functions of the nerves, for he divided them into nerves of sensation and nerves of motion. The animal spirits he seated in the brain, the vital spirits in the heart.

Claudius Galen was a Roman, born in Pergamus, in the year 131 A.D., a famous centre of learning at that time, next in importance to Alexandria across the Mediterranean Sea. He studied Aristotle and Plato under his father and Guius, the latter being a Stoic and an Epicurean. He pursued medicine in Alexandria, Smyrna, and elsewhere. He then tried to reform medical theory and practice, and wrote extensively. He appears to have been at first an Eclectic, then became a Dogmatic, then revived and remodelled the teachings of Hippocrates and added to them. He also excelled as an anatomist, and is the first we have account of who specially treated nervous diseases.

Galen was decidedly opposed to the Epicureans. "After a burst of indignation against all who would place their supreme good in the gratification of their own will, he exclaimed: 'Why should I waste words on such men? Others of nobler views might well censure me for thus perverting the sacred attribute of speech, which ought to be reserved for composing hymns in adoration of the Author of our being.' I hold true piety to consist not in sacrificing to Him hecatombs of bulls, or in burning incense of cassia, or offering fragrant ointments to His honor, but rather learning for myself, and in teaching to others something of His wisdom, goodness, and power. All are supplied by His goodness, and all are supported by His bounty. On this account it becomes us to celebrate His goodness with hymns of praise, and observe the evidences we have of His almighty power."*

"The most of those works upon which Galen's fame reposes were written after his recall to Rome. Some portion of his anatomical and other writings were consumed in the conflagration of the Temple of Peace and the destruction of his own dwelling near by it. Yet we have now in print eighty-two treatises undisputed, and eighteen commentaries on Hippocrates, to which should be added nearly forty treatises or fragments of treatises still extant in manuscript. His works now wholly lost

* "De Usu Partium," lib. iii., c. x., vol. iii., p. 237; P. Watson's "Medical Profession," p. 161.

are supposed to amount to one hundred and sixty, about fifty of which were on medical subjects, too numerous to mention here."

In the first of the works cited above, we are told of the few occasions enjoyed by the ancients for acquiring anything like correct anatomical knowledge. The first five book so fhis treatise on anatomy are occupied in describing the muscles, many of which are mentioned for the first time. He also speaks of the blood-vessels. In the sixth book are described the organs of digestion. In the seventh, the heart ; in the eighth, the respiratory organs; in the ninth, the brain and spinal marrow ; and in six remaining books, which have perished, he treated of the eye, the tongue, the pharynx, the larynx, the os hyoides, the history of the arteries and veins, the cerebral and spinal nerves, and the organs of generation. Many facts found in this work have been claimed to be the discoveries of later times.

"Soon after his return to Rome he wrote *De Usu Partium*, also replete with physiological opinions and on final causes, the author's main object being to disprove the doctrines of Epicurus, and demonstrate the existence of a superintending Providence, from the wonderful adaptation of means to ends in the organization of the human frame. In this treatise is found ,that eloquent exposition of the powers and uses of the human hand. And here, too, are found those hymns to the Deity and other pious expressions so worthy of the philosopher, moralist, and physician. This work is in seventeen books and has been preserved entire.

"The treatise *De Locis Affectis*, in six books, was the work of his maturer years. In this, with wonderful sagacity, he points out every part of the body subject to pains, convulsions, paralysis, or other symptoms, and to which our attention should be directed for investigating into the causes of disease. This valuable treatise, which is occupied mainly with pathology and symptomatology, the learned Dr. Haller held in higher estimation than any other of Galen's works. Already, however, Aretæus, in the first century, had written on nervous affections, and left two books upon ' The Lethargies.'

" The treatise *Ars Medica* was for many centuries the text-book upon which the students of Salernum and other medical schools of the middle ages were examined before receiving permission to practice. Commencing with the definition of medicine, it treats of the signs of health, of the temperaments generally, and of their influences on special organs in health and in disease. It next treats of the signs of disease, general and local, of prognostic indications, of causes of disease, of the means of preventing disease and preserving health, and of restoring it when disordered, thus furnishing in small compass an exposition of the whole of Galen's system of medicine.

" The *Methodus Medendi*, in fourteen books, was the work of his old age, and was held by his followers in nearly the same estimation as the 'Ars Medica.' The two books, on the same subject, addressed to Glauco, treat mostly of generalities. More information, it has been said, may be obtained from this work than from the whole medical literature of the Arabians.

" His original investigations were chiefly in the department of anatomy. In this he made many discoveries, mostly in the muscular system. He was the first to describe the popliteal muscle, the platysma myoides, the sterno and thyro hyoideus, and probably many others. In the blood circulation he was not much in advance of the early Alexandrians. Like them, he placed the origin

of the veins in the liver, of the arteries in the heart. He
was familiar with the anastomoses of the two orders of
vessels, the arteries and veins, but he traces the current
of blood from the liver, the supposed fountain of the
venous portion, into the right ventricle of the heart, and
then through the pulmonary vessels to the left ventricle.
He was acquainted with the uses of the opening between
the right and left auricle, but in other respects his ideas
of the course and distribution of the blood were confused
and incorrect. He appears to have made several discoveries in the nervous system : he points out the tubercula
quadrigemina, the corpus callosum, and septum lucidum ;
he derives the nerves of sensation from the brain, those
of motion from the spinal marrow, and to some nerves
he assigns both sentient and motive power; he denies
the decussation of the optic nerves, but admits their
junction at the commissure ; he describes the par vagum
and its connections with the sympathetic nerve. Certain organs, as the heart and blood-vessels, he supposed
to be destitute of nerves, and hence devoid of sensibility.
He took much pains to determine the structure and development of the human fœtus. He allowed no occasion
to escape for impressing upon his students the importance of anatomical knowledge.

"In respect to physiology, he speaks of the living
body as a unit, though constituted of parts or organs,
simple and compound, of humors, and of spirits. With
the old opinion, he maintained that all the parts, by
which he means the material structures, whether simple
or compound, are constituted of the four primitive elements, fire, air, earth, and water ; from which are derived
the four corresponding qualities, the hot, the cold, the
dry, and the humid. He also enumerates four humors :
the blood, which is red, hot, and moist ; the phlegm,

which is white, cold, and humid; the bile, which is yellow, hot, and dry; and the melancholia or atrabile, which is black, cold, and dry. These two latter he holds to be partly excrementitious. From the combination of these elements and their respective qualities results the complexion, or chrasis, of each part or texture of the body. The preponderance of one or the other of the four humors gives the corresponding temperament.

"As did the Peripatetics, he attributed the essential phenomena of life to certain *occult forces*, inherent in the several parts or organs. These forces he divided into the vital, the animal, and the natural. The seat of the first is in the heart; of the second, in the brain; and of the third, in the liver. But above all these forces he admits, with Hippocrates, the presiding and ruling influence of Nature,—a word by which the ancients meant vitality, vital force, organic force, or principle of life. The spirits, under the name of pneuma, he also divided into the vital, the animal, and the natural, corresponding with the respective forces by which the functions of the body are performed. The natural, or least attenuated of the spirits, are evolved from the blood in the liver, the organ in which the blood itself is first elaborated. The spirits thus conducted with the blood to the lungs, and there exhaling certain impurities, and combining with the respired air, the natural are converted into the vital spirits; and then passing to the brain, the vital becomes still further attenuated and converted into the animal spirits.

"The functions, he said, are also of three orders, the vital, the animal, and the natural. To the first of these belong the action of the heart and arteries, also the passion of anger and revenge; to the second belong the intellectual powers, intelligence and sensibility; to the third belong the functions of nutrition, muscular action, and generation ; and these functions he further divided into the external and internal. By the intervention of the pneuma the vital force produces the pulsation of the heart and arteries. He says the blood is cooled, the pneuma is relieved of its fuliginous particles, and the blood is endowed with the vital force by the process of respiration ; and that this process is affected by means of the diaphragm and intercostal muscles.

" Galen held that the brain is the seat of the rational mind; the heart is the seat of courage, and the angry passions ; the liver is the seat of desire. By the internal pulsation of the brain, the pneuma of the ventricles of the brain is engendered, and in these ventricles the functions of the mind are executed. The passage of the vital spirits from all parts toward the brain, where they acquire new qualities, explains in what way the mind is influenced by the body. But it is not clear whether he looked upon the mind as an entity, or as a mere result of organic action. From the brain, by the energy of the nerves, sensibility and motive power are diffused throughout the body : but special forces, subordinate to the mind, preside over the function of special sense. That the brain in the performance of its functions exudes a pituitous humor, which is discharged through the foramina in the cribriform plate of the ethmoid bone, and escapes through the throat and nostrils. And that every organ has its own peculiar force of attraction, retention, and expulsion, and some of them similar to the forces more recently termed endosmosis and exosmosis. Thus by its own peculiar power the stomach attracts the aliment, retains it, concocts it, or expels it. From this, it is seen that his physiological opinions were partly hypothetical.

Galen held that "health is maintained by supplying similar with similar, whilst disease is overcome by opposing contraries to contraries." These two propositions furnish the key to his whole system of hygiene and therapeutics. Galen wrote no work expressly on the practice of medicine, but he has left a complete code of medical science, the only complete code of which we read among the ancients. In prognosis he was remarkably skilful. He treated of indications and contra-indications, and demonstrated the superiority of the practice of the rationalists over the school of the empirics. The art of medicine in Galen's time consisted mostly in devising, or applying, particular remedies to particular diseases, each having its own specific.

"Galen held himself as superior to Hippocrates, for he assumed that the latter had only commenced what he had carried to completion. Though he did endorse Hippocrates in many things, yet they were quite unlike in others. Hippocrates wrote with the terseness of a philosopher; while Galen wrote with flowing rhetoric, and adorning his discourses with criticism, anecdote, sarcasm, and boasting. Hippocrates seemed to draw his inspiration from the workings of nature." While Galen lies buried in the accumulations of his own superabundant labor, Hippocrates, as says Dr. John Watson, "lives to be studied with as much edification by the physicians of the present day, as by his own immediate disciples," in his own time.

When Galen died, about the year 200 A.D., all that was aggressive in medical science at Rome then ceased. And as it decayed there it again revived at Alexandria, where it continued until that city was destroyed by the Saracens. After the removal of the imperial court of Rome to Constantinople the scholars that still lingered about the pal-

ace of the Cæsars were attracted to the new capitol, so
that but a few authors of importance appeared after this
at Rome.   One of the first after this was Quintus Serenus
Sammonicus; but though a writer, he was mostly famous
for his trait of superstition.   He held a great veneration
for the numbers three, seven, and nine.   He recommended the use of amulets, one of which he calls the
Abracadaba.   This amulet consisted in writing this word
on cloth or parchment, in full, on the upper line, and
then on every succeeding line omitting a single letter
until the initial letter was at length the only one remaining to be written on the last line, this making, by
means of these letters so written, a triangular figure,
which was to be suspended by a string or chain, and
worn around the neck as a preventive against fevers.

Then appeared Theodore Priscian, the pupil of Vindi-
cian, physician to Valentinian ; also Marcellus Empiricus,
but nearer the close of the fourth century, some time
after the struggle for the supremacy of Christianity in
the Roman empire had been crowned with complete success.   He was a Gaul, a native of Bordeaux, and was
probably a Christian.   He speaks of his contemporaries
Siburius, Eutropius, and Ausonius ; also of Greek and
Latin physicians, as Pliny, Cornelius, Celsus, Appolinaris,
and of Renatus, the author of veterinary medicine, in
four books.

Among the Greeks after the death of Galen there were
Cæsarius, Oribasus, and Nemesius, which was between
A.D. 350 and the death of the elder Theodosius A.D. 395,
about the period of the division of the Roman empire.
" Cæsarius was the younger brother of St. Gregory Nazi-
anzen, archbishop of Constantinople, and was the earliest
Christian physician of distinction at the imperial court.*

* See Watson's " Medical Profession," p. 179.

On the completion of their elementary studies, the two brothers, Gregory and Cæsarius, departed on the same day from their father's house, the one to complete his philosophical course at Athens, still the principal seat of Grecian arts, and the other to pursue his medical studies at Alexandria. After five years' study he commenced practice, and rose to high distinction in his native place, Nazianzus, a town of Cappadocia. He then removed to Byzantium, and under the patronage of Constans, Cæsarius soon rose to the senatorial rank, and was appointed archiater, that is, physician to the emperor. Julian, who had formerly been a fellow-student at Athens, and who succeeded to the throne A.D. 363, urged him to return to the ancient religion of the state ; and though he persistently refused, he was retained at the imperial court. He was a writer in both medicine and theology."

There prevailed at that time in most of the young students at Athens a complete sophistic furor. They all canvassed for their master ; for it was not the custom to attend different lectures at the same time, but each one, as a rule, attached himself to one teacher. The poor students especially lent themselves to this business of recruiting, since they got exemption from class payment. An honest youth could scarcely set his foot on Attic ground without being already claimed by the adherents of a party. They wrangled, they struggled, they threw themselves around him ; and he might be torn quite away from the teacher he came to attend.

Oribasius was a native of Pergamus, or Sardis, and received his education under Zeno of Cyprus, the most famous medical professor of the fourth century, who taught first at Sardis and afterwards at Alexandria. "If Cæsarius was the earliest of the Christian, so was Oribasius among the latest of the pagan medical writers.

He ranked with the philosophers of the times, and his influence was such that Julian was largely indebted to him for his accession to the throne." He was one of the four friends who were permitted to accompany the emperor into Gaul. At his request he undertook a journey to Delphos, and received from that oracle the memorable response, that "hereafter the oracles are mute." He accompanied Julian into Persia, where he attended the wounded emperor in his dying hours after the defeat of the imperial forces. Familiar with the literature of the medical profession from the earliest ages, he undertook, at the request of Julian, the compilation of all that was valuable in previous works on medicine.

Constantine ruled from A.D. 306 till 337, under whom Christianity first became the national religion of the whole Roman empire. The edict of this emperor for the closure of the Asclepions, as well as other remaining temples of pagan worship, led at once to the establishment of hospitals and other charitable institutions under the care of the church.* Before this time the religious and moral instruction of the flocks was but a part of the duties of the clergy, for they had charge of the care of the orphans, the widows, the poor, the friendless, the needy and suffering: for while Christianity was obtaining the ascendancy, the clergy were gradually acquiring the control of all that related to the physical and social welfare of the people. This custom originated probably with the local organization of the churches in the time of the apostles.

Helena, the mother of the emperor, devoted all her pious energies to the founding of churches and other benevolent institutions at Constantinople, at Jerusalem, and in other places.† In every part of the empire this

* Eusebius, book x., chap. vi.
† Evagrius. "Eccles. History," book iv.

noble example was followed by other ladies of wealth and influence, especially until in the succeeding reign of Gallus Cæsar, the elder brother of Julian. The groves of Daphne, in the neighborhood of Antioch, once sacred to Apollo, were in A.D. 357 dedicated to the church. The grounds, within which formerly stood the magnificent temple of Apollo Daphnæus, were occupied by a hospital for the sick.* After this, the emperor Valens presented the most beautiful lands near Cæsarea to Archbishop Basil, for the poor who were also sick. As early as in 373 Basil had already organized an immense hospital at Cæsarea, called the Basilides, which was often mentioned as among the then wonders of the world, so numerous were the sick poor, and so admirable was the care of them. The most illustrious ladies of the empire participated in these offices of mercy. At Constantinople the Empress Flacilla, wife of the elder Theodosius, in 380, was active in watching and nursing the wounded and the mutilated: she made ready their food, and often carried their dishes to them, and visited others at their own homes, waiting upon them herself, and supplying their wants.

In the different orders of the clergy under the empire there were the Brephotrophi, who had the care of foundlings. The Ptochotrophi had the charge of the poor, and the Nosocomi had the care of the sick. Then there were the Perabolani, who, to the number of six hundred, served under the bishop at Alexandria, whose duty it was to care for the sick in time of pestilence, and these latter were uneducated laymen. The first three orders mentioned above were obliged to study medicine to a certain short degree.

Aëtius, of Amida in Mesopotamia, a learned physician

* Theodoret, "History of the Church," book i., chapter xviii.

and of the Christian faith, is supposed to have lived about the middle of the sixth century. He practised in Constantinople and was a famous surgeon, yet he had much to do with amulets and incantations; also using Scriptural expressions in the preparation of his medicaments, for the purpose of imparting to them greater efficacy.

Alexander, surnamed Trallianses, from his native city, Tralles in Lydia, appeared shortly after Aëtius. He was of a talented family, and it was one of his brothers, Authenias, who was employed by Justinian as an architect in building the cathedral of St. Sophia in 532, which still stands prominent among the ornaments of Constantinople. Alexander settled and practised in Rome. His work on the Art of Medicine is received next in importance after Hippocrates and Galen.

Gibbon says the reign of Justinian was one of much vigor from 527 to 565, and marked with successes against the overwhelming inroads of barbarism. Italy was for a season delivered from the Goths, and the two portions of the Roman empire again united. The whole body of Roman law was at this time revised and reduced to system. The virgin schools of Constantinople were excellent and rising into notice. From the reign of Justinian to the downfall of Alexandria from the overrunning of the Mahometans but little was done in medical literature excepting by Ahrun of Alexandria, who was the first to write on small-pox. Paulus Ægineta was educated at Alexandria, and perhaps taught in that school about 640 A.D., to the period of its final subjugation by the Saracens. The only other writer among the Greeks after the time of Paulus was Actuarius, who is worthy of rank among the ancient classics of the profession.

Some nervous affections must have been observed from

earliest times.   Herodotus* mentions nervous disorders
as the affliction of the Scythians 550 years before the
Christian era.   He says, "The Scythians, having obtained
the entire possession of Asia, advanced towards Egypt.
The king of Egypt met them in Palestine of Syria, and
by presents and importunity prevailed on them to return.   The Scythians, on their march homeward, came to
Ascalon, a Syrian city; the greater part of the army
passed through without molesting it, but some of them,
remaining behind, plundered the temple of the celestial
Venus.   Of all the sacred buildings erected to this goddess, this, according to my authorities, was far the most
ancient.   The Cyprians themselves acknowledge that
their temple was built after the model of this, and that
of Cythera was constructed by certain Phœnicians, who
came from this part of Syria.   On the Scythians who
plundered this temple, and indeed on all their posterity,
the deity entailed a fatal punishment: they were afflicted with the *female disease*" (hysteria, or nervous disorders).   "The Scythians themselves confess that their
countrymen suffer this malady in consequence of the
above crime.   Their condition also may still be seen by
those who visit Scythia, where they are called *Enarcæ*,
or effeminate."

* Herodotus, book i., clio cv.

4

CHAPTER II.

THE DARK AGES—MAHOMETANISM IN EUROPE; MEDICINE, CHRISTIANITY, AND LAW, BLIGHTED.

ABOUT the year 700 A.D. the art and science of medicine was greatly affected by the inroads of Mahometanism into Europe. The story is here made very short. The Arabian Prophet, Mahomet, was born at Mecca in Arabia in 570. At forty years of age he began to proclaim his doctrine. His motto was, " There is no God but God, and Mahomet is his Prophet."

" The people and the elders of that city growing weary of him resolved to put him to death. He fled from Mecca by night through the palm groves to Yatreb, in 622. This is the epoch of the *Hegira*, from which the Arabs compute time. In 630 Mahomet returned to Mecca already accepted as Prophet and Prince. He is said to have purified the Kaaba Temple and destroyed three hundred and sixty idols. He then decreed that no infidel—no disbeliever in him—should enter the holy city. When Mecca had become obedient, and all Arabia paid him reverence, he next commanded that *Islamism* be carried into every country, and that all nations be made to unite in it, by faith or by force of arms." * His followers, inspired with this fearful propagandism, soon overran, by means of the sword, all northern Africa and southern Europe. But the Mahometans did not continue united, and this caused the founding of two great uni-

* R. Laberton, p. 104.

versities of learning, where medicine was long taught;
the one in Bagdad, the other in Cordova in Spain, the
two capitals of the two caliphs, the eastern and the
western.

The Fatimates are the descendants of Fatima, the
daughter of Mahomet, and her cousin and husband
Ali, who in 656 was proclaimed the fourth caliph. The
three before him were Abubekar, Omar, and Othman.
Ali was slain in 680, and left two sons by Fatima. Part
of the sect followed the one, and the other part followed
the other. The line of one entirely run out ; but Hosein,
the last of them, is believed by a portion of the faithful
to be still alive in concealment, and will appear as sovereign in the end of time. Such are the Persians. The
other sect, called the Turks, or Sunnites, fill the Ottoman empire, and are execrated by the former as enemies
of the great Ali. The will of the High Pontiff of Is-
lam finally became superior in power from the Indies
on the east to the Atlantic Ocean on the west.

During the eighth century was the greatest extent of
the caliphate. The house of Ommiah, under whom the
conquests throughout Arabia, Egypt, Africa, and Europe
had been made, were finally hunted down by the other
party, so that but a single youth of this doomed race
or branch of Mahomet's family escaped from destruction. He finally found his way to Spain, where parti-
sans rallied to him, and made him sovereign of the
country. From him came the eminent Abderrahman
line of emirs, caliphs of Cordova. While in the east, after
Abul Abbas had been established as caliph on the throne
at Damascus, the original capital of the caliph was
transferred to Bagdad, on the river Tigris, where the
Abassides ruled over Mahometan Asia for more than
five centuries, until extinguished by the siege of the

Mongols, who finally stormed Bagdad, the only city in the east at that time left in the possession of the caliphs, and for several days the streets of the city were deluged in blood. Here, the fifty-sixth and last caliph, was killed in 1258. From the early year of 750 we see that Mahometan history lost its unity, for the empire was permanently divided, and remains so till this day.

At Cordova, while the caliphate was there, arose the most learned and extensive university in all Europe in that age. While in Bagdad, in the far east, was another great school of philosophy and learning, where seven thousand medical students, besides others, attended at a time. This appears to have not arisen so much from the love of medical lore by the Arabs, as it was rather to establish their priesthood, and to embellish Mahometanism, and fortify it against the advance of Christianity. Colleges were also established at Seville and Toledo in Spain. These extended to the thirteenth century. Then there flourished a university at Bologne; also others at Padua, and Naples, and Salerno in Italy, where arose the anatomist Mondino, who studied the nervous system in particular; and about 1500 A.D. appeared Alexander Achillini of Bologne, who studied the brain.

The next largest medical school, *in that age*, of the world was at Bagdad, built in 765 by the Caliph Almansur, on the Tigris, east of Jerusalem. There were numerous professors, and about an average of seven thousand students resorted to them yearly from all parts of the surrounding nations, east and west, north and south. As those people in the east change but little, this great school was probably much like the present great, but little known, university of the Mahometans at Azar, in Egypt. Here, at the present time, is said to be the centre and chief source of Mahometan learning, including the medical, for all Moslem nations, and has been for the past thousand years. Only the Koran, and its prescribed literature, is taught at this great Moslem university. There each of the three hundred and fourteen professors selects his own specialty, and each student his teachers. Among these departments are taught the history, theology, and jurisprudence of the Koran.

The University building at Azar, an immense quadrangle, with a spacious court in its centre, has around its front and sides a continuous portico, which is supported by four hundred large marble pillars, gathered mostly from very ancient heathen temples. At present there are nearly seven thousand students in attendance, representing all lands possessing the Mahometan faith, and they learn only what their ancestors knew and their profoundly Mahometan professors choose to teach.

The healing art thus became lifeless and comparatively useless during all the dreadful epidemics and numberless sicknesses and sufferings that occurred through the long dark ages! The Latin Church itself became lifeless to the millions in Europe, though, according to history, it continued throughout all the ten persecutions to still exist in purity; but when it came forth from the Catacombs to take possession of the Basilicas a change for the worst was soon observed.* It struggled for three hundred years against paganism in the Roman empire, then it struggled for three hundred years against Arianism; and after all this it had to conquer heathen Germany and the old Scandinavian and Hungarian oppressors, besides braving the opposing tidal-wave of Mahometanism that ultimately overran all southern Europe. At last it had to conquer the licentious-

* Laberton.

ness and other lawlessness which prosperity had in-
dulged in her own ranks. So that the tenth century
marks the darkest period in the history of the Christian
Church; and this was equally true of the medical profession.

Near the close of this time, about 990 A.D., the Rom-
ish Church under the popes had not only lost all commanding authority, but could not even maintain outward
decency; for during this time arose into power the infa-
mous Theodora, with her daughters, who, in the strong
language of contemporary historians, disposed for many
years of the papal tiara; and not content with disgracing
by their own licentious lives the chief city of Christendom, they actually placed their profligate paramours or
base-born sons in the chair of St. Peter's! Astonishing
as it may now appear, these twin sisters, Medicine and
true Christianity, adapted as they were to the constant
highest good of mankind here and hereafter, became
blighted, and even perverted, and thus nearly submerged
they drifted on aimlessly through the ages.

When alluding to the Royal Family of Portugal, the
writer Joaquin Antonio De Macedo stated that "after
the conquest of the peninsula of Spain and Portugal by
the Arabs, at the beginning of the eighth century (in
711), the Christians were reduced to the abject condition
of slaves; their former distinctions ceased, and they were
all made equals in misfortune, and were driven to take
refuge in the mountains of Asturias. After several centuries of sufferings the Christians succeeded in shaking off
the hated Mussulman yoke, and established the monarchy
of Asturias. The first distinctions which appeared after
that were the *escudeiros*, applied to those who fought
with sword and shield (escudo), *cavalleiros*, those who
possessed a horse (cavallo), and *ricos-homens* (literally, rich

men), who had acquired fortunes at the expense of the enemy. Subsequently, when the kingdom of Asturias merged into the more extensive monarchy of Leon, titles of nobility were for the first time introduced at court, and they consisted of *ricos-homens, infancocs,* and *vassallos.* Affonso Henriques, when he founded the Portuguese monarchy, followed the example of Spain, and introduced the same titles." The University of Lisbon was not founded until 1289. The medical department was, and is, connected with a great hospital of about 900 beds for the sick and for surgical cases.

" *Mahomet the Second,* Emperor of the Turks, surnamed the Great, was born about eight hundred years after Mahomet the prophet, at Adrianople, in 1430, and succeeded his father, Amurath the Second, in 1451. It was under his reign that the Turkish Empire in Europe was raised upon the ruins of the Greek. His ancestor, Ottoman, or Othman, who rose from the rank of a common soldier, became a general of a sultan of Iconium, and laid the foundation of the grandeur of his house. He possessed himself of a part of Bythinia, and of Cappadocia. Then his son Orchan added to these possessions Mysia, Caria, and all the provinces that extend towards the Hellespont. Amurath I. afterwards subdued the whole of Asia Minor, passed the Bosphorus in 1355, and took Adrianople, the second city in the empire, where he fixed his residence.*

" In 1429 Amurath II. resumed the project of his ancestors, and, having been successful in Asia, crossed the Hellespont, possessed himself of Thessalonia, poured his troops into every part of the Grecian empire, destroyed, in 1444, the Christian army in Hungary, and gave an emperor to Constantinople in 1448. From this race of warriors sprang Mahomet II., who, in cruelty, bravery, and

* Historical Library.

conquest, surpassed them all.   He was twice called to the
throne during the reign of his father, but resigned it in
favor of his father.   In 1451 he a third time took the sceptre, and resolved then to take possession of Constantinople, which his father had deferred.   He invested that city
with 300,000 men, and by a considerable fleet ; it was
carried by assault in May, 1453.   Constantine Dragaces,
the last Turkish emperor, contended against his unhappy
destiny with the courage of a hero.   Betrayed by his subjects, abandoned by Europe, he perished, sword in hand,
on that memorable day which eclipsed at once the liberty
of the Greeks, the name of the Cæsars, and the glory of
an empire which had subsisted for fifteen centuries.

"During the pillage of Constantinople, and all its attendant horrors, a pasha conducted to Mahomet a young
princess, named Irene, whom her innocence and her beauty
had saved from the general carnage.   On seeing the de-
stroyer of her country she burst into tears, and fell at his
feet.   Her youth, her anguish, and her tears augmented
her attractions in his eyes.   Mahomet for a moment
contemplated her beauty, then dragged the victim to his
palace, and for three days delivered himself to the gratification of his pleasure.   His janissaries, indignant at his
conduct, began to murmur ; a vizier ventured even to reproach his sensuality.   Mahomet immediately ordered
his captive to be brought before him, and, in the presence
of his officers, severed her head from her body, saying,
"It is thus that Mahomet releases himself from love."
Three days afterwards he made his triumphal entry into
the city, distributed rewards to the conquerors and his
treasures among the vanquished, while he installed himself a patriarch.

"Mahomet II. thus rendered himself notorious by rav-
aging the earth.   During thirty-one years he subdued two

empires, twelve kingdoms, and two hundred cities, meeting little or no resistance from the Christian princes, who
could not unite to oppose so formidable an enemy."
Nor need we wonder, since such is history, that there
should have been made up the awful story of a " Blue
Beard," who is said to have tormented and destroyed so
many beautiful and innocent ones; a story that is now
become familiar throughout all Christendom.

> " There was a poor *parson* of the town
> But rich he was of holy thought and work.
>
> .   .   .   .
>
> Wide was his parish, and houses far asunder
> But he ne left not, for no rain nor thunder
> In sickness and in mischief to visit
> The farthest in his parish, much or lit,
> Upon his feet,--and in his hands a staff
> This noble ensample to his sheep he gaf
> That first he wrought and after that he taught.
>
> .   .   .   .
>
> He was to sinful men not dispiteous
> Nor of his speech dangerous nor deign,
> But in his teaching discreet and benign.
> To drawen folk to heaven with fairness
> By good ensample was his business.
>
> .   .   .   .
>
> He waited after no pomp nor reverence
> Nor maked him a spiced conscience,
> But Christes love and His apostles twelve
> He taught, and first he followed it himself."
>
> *Canterbury Tale, Prologue.*

Here we cannot but notice that parallel line in history
—I refer to sacred story—termed " The Scriptures," or
" The Word of God." These two prominent lines, or
as we may say, *two strands* of the warp in the loom of
time, God's Word and Medicine, are probably the most
important of all the threads of ancient story. In some

respects these two seem much alike, as their aim and
tendency are for the best good of the human race, each
for sanitary and healing effects; yet the one is greater
than the other. Moreover they are quite unlike in their
source and perfection. The sacred Word came from our
great Creator, and was in all fundamental respects perfect from the beginning, bringing with it only truths,
laws, and commandments, and a scheme of everlasting
salvation, all-sufficient for each and every individual
spirit, which has only developed in the course of time.
These two were brought near together in the person of
Christ.

Medicine began indefinite and imperfect, containing
an admixture of truths and errors that have needed correcting and improving in every age, nor needs it less to-
day. The minister and the missionary has a sure word
of gospel from which to preach and prescribe, though
the same old story, yet always fresh, true, and *sufficient*,
if dispensed with skill and the aid of the Holy Spirit, as
promised; for he can prescribe a practice that is infallible in every case, if only *believed* and *lived;* while the
physician, like every other philosopher and scientist, is
obliged to work on and change his base from time to
time as knowledge and means increase, and so feel his
way on to more nearly complete and uniform methods
and rules, yet never infallible.

Those royal writings of the Bible were first arranged,
under authority, by Moses, and then, after more than a
thousand years, they, with additions, were collated by
Nehemiah, and then again by Ezra, about a century before Hippocrates. They were again re-written for the
Alexandrian Library in Egypt, about two hundred years
before the Christian era. The Samaritans carefully preserved a complete copy, and the Jews have always kept

another copy with a nation's care and pride all along the ages ; so that, unlike medical writings, they were preserved more select and sacred, pure and perfect. In fact, the word of God was bound up by itself as a precious scroll, in the preservation of which all nations were interested. The Bible was completed, with the addition of the gospels, epistles, and the Revelation of St. John, about 100 A.D. All culture in all civilized nations now recognize its heavenly influence, so that it is indeed *The Book* for every individual to know, respect, and delight to read.

An Example of the Sphere and Service of Physicians in very Ancient Times.

The following scene and conversation, a word-picture, mostly taken from "The Egyptian Princess," may concisely illustrate the ordinary daily life of the physician and people of ancient Egypt about B.C. 528. This was the actual occasion when the famous Crœsus, of former fabulous wealth, visited Amasis, the reigning Pharaoh ; which also accords with the account by Herodotus.* The writer opens the scene by stating, "It is not an approaching army, but a grand embassy from Cambyses, the ruler of that most powerful kingdom, Persia."

"Said Phanes, 'At Samos I heard that they had already reached Miletus, and in a few days they will be here. Some of the king's own relatives are among the number; the aged Crœsus, king of Lydia, too ; we shall behold a marvellous splendor and magnificence ! Nobody knows the object of their coming, but it is supposed that Cambyses wishes to conclude an alliance with Amasis ; indeed, some say the king solicits the hand of Pharaoh's

* Dr. G. Ebers, " Egyptian Princess," p. 14.

daughter.' 'An alliance?' asked Phanes, with an
incredible shrug of the shoulders. 'Why, the Persians
are rulers over one-half of the world already. All the
great powers have submitted to their sceptre ; Egypt
and our own mother country, Hellas, are the only two
that have been spared by the great conqueror.'

"'You forget India, with its wealth of gold, and the
great migratory nations of Asia,' answered Kallias.
'And you forget, moreover, that an empire composed,
like Persia, of some seventy nations or tribes of different
languages and customs, bears the seeds of discord ever
within itself, and must, therefore, guard against the chance
of foreign attack, at least while the bulk of the army be
absent, also lest single provinces should seize the opportunity and revolt from their allegiance. Whatever the
intentions of the envoys may be, they will be here within
three days.'

" The old King Amasis received the Persian embassy
shortly after their arrival, with all the amiability and
kindness peculiar to him. Soon after he was seen walking with Crœsus in the palace gardens. 'Does happiness
consist in possessions?' asked Crœsus, the ex-king, whose
sumptuous palace had been in Sardis. 'Is happiness
itself a thing to be possessed? Nay, by no means! It
is nothing but a feeling, a sensation, which is vouchsafed
more often to the needy than to the mighty. The clear
sight of the latter becomes dazzled by the glittering treasures, and they cannot but suffer continual humiliation ;
because, conscious of possessing power to obtain much,
they wage an eager war for all, and are therein contin-
ually defeated, and consequently unhappy.'

"'And yet,' said Amasis, 'Death has for us too his
terrors, and we do all in our power to evade his grasp.
Our physicians would not be celebrated and so esteemed

as they are if we did not believe that their skill could
prolong our earthly existence. This reminds me of the
oculist Nebenchari, whom I sent to Susa, to the king.
Does he maintain his reputation? Is the king content
with him?'

" ' Very much so,' answered Crœsus. ' He has been of
use to very many of the blind ; but the king's mother is,
alas! still sightless. It was Nebenchari who first spoke
to Cambyses of the charms of thy daughter Tachot.
But we deplore that he understands the diseases of the
eye alone. When the Princess Atossa lay ill of a fever
he was not to be induced to bestow a word of counsel.'

" ' That is very natural,' replied Amasis the king. ' Our
physicians individually are only permitted to treat one
part of the body. We have aurists, dentists, oculists, and
surgeons for fractures of the bones, and other physicians
for internal diseases. By the ancient priestly law (for
all physicians are priests) a dentist is not allowed to
treat a deaf person, nor a surgeon to treat a patient with
disease of the bowels ; though he should have a first-rate
knowledge of internal complaints. This law aims at
securing a great degree of thorough practical knowledge ;
an aim, indeed, pursued by the priests (to whose caste
the given physician belongs) with a most praiseworthy
earnestness, not only in medicine, but in all branches of
science. Yonder is the house of the high-priest Nei-
thotep, whose knowledge of astronomy and geometry
was so highly praised even by Pythagoras. It is, you
see, there next to the porch leading to the temple
of the goddess Neith. Would I could show you the
sacred grove, with its magnificent trees, the splendid
pillars of the temple, with capitals modelled from the
lotus flower, and the colossal chapel, which I caused to
be wrought from a single piece of granite, as an offering

to the goddess; but, alas! entrance is strictly refused to strangers by the priests.

"'Many prominent in the healing arts have been brought up in the house of Seti.* But few remain after passing the examination for the degree of Scribe. The most gifted were sent to Heliopolis, where flourished, in the great "Hall of the Ancients," the most celebrated medical faculty in those times; whence they returned to Thebes, endowed with the highest honors in surgery, in ocular treatment, or in some other branch of the profession, and then became physicians to the king, or made a living by teaching and by being called in to consult on serious cases. Most of the doctors lived on the east side of the Nile, in Thebes proper, and even in private houses with their families, but each was attached to a priestly college. Who ever required a physician, sent for him, not to his own house, but to the temple. There a statement was required of the messenger concerning the complaint from which the sick person was suffering, and then it was left with "the principal of the medical staff of the sanctuary" to select that master of the healing art whose special knowledge and experience qualified him to be best suited for the treatment of the case.'"

"Like all priests, the physicians and surgeons lived on the income which came to them from their landed prop-

* What is here stated with regard to the medical schools refers to a period as far back as fifteen centuries before the Christian era, and is principally derived from the medical writings of the Egyptians themselves, among which the "Ebers Papyrus" holds the first place, those in Berlin the second, and a hierati MS. in London, which, like the first mentioned, has come down from the eighteenth dynasty. See also Herodotus ii., p. 34; Diodorus i., p. 82. Among the six hermetic books of medicine mentioned by Clement of Alexandria was one devoted to surgical instruments; otherwise the badly set fractures found in some of the mummies, do little honor to the Egyptian surgeons, unless these cases or bones received no surgery.

erty, from the gifts of the king, the contributions of the laity, and their regular share of the state revenues. They expected no honorarum from their patients. Though, according to every indication, the former Egyptian medical man had much real knowledge and skill, yet, standing by the bed of sickness as the ordained servant of the Divinity, it is but natural he should not be satisfied with a simple rational treatment of the sufferer, but should think he could not dispense with the mystical, and efforts of prayers and vows."

So it appears that, " by law," medicine was then practised only in specialties; for all physicians and surgeons were specialists. Even in those very early times, as noticed also in the Epos of Pentaur, and by historians, when they dissected animals only or mostly, and " *tested* " their medicinal herbs and roots on both animals and man, they discovered the nature of strychnine or nux vomica, the hydrocyanic acid or " poison of the peach," and scammony, eleterium or the squirting cucumber, conium, aconite, aloes, senna, and manna, etc. Very fundamental principles were put on record, too, in those times, and one thousand years afterwards some of them were repeated by Hippocrates. As we now read and re-read them, they appear like real diamonds among the trash and dirt. Some of those very ideas we still accept; and thus a multitude of important facts were discovered along the ages, and have been gleaned for our use by the old and still older writers and teachers of the different sects and schools before two hundred years ago.

Throughout all those ages little or nothing is said of the origin or nature, or even the existence, of nervous affections, excepting hemiplegia, paraplegia, and epilepsy. As to diseases in general, but little was known; materia medica and pharmacy were sadly behind surgery and obstetrics. Yet what facts did accumulate from time to time have been for the fathers, and are now for us, of inestimable value, when joined with our newly acquired chemistry, physiology, and pathology and histology, and a better understanding of nature in disease, as well as in health. Moreover, not a sect now exists, not a school of old dogmatic or empirical medicine survives, that is recognized by the great learned body of regular physicians. Not one " old-school creed " of any of the former sects in medicine, so popular ages ago, is anywhere respected, or even tolerated. All have been tested and sifted, and all the valuable morsels have been selected out to enrich our present didactics; while the immense refuse of doctrines, speculations, and creeds have been completely blown to the winds forever.

CHAPTER III.

THE MEDICAL PROFESSION ABOUT TWO HUNDRED YEARS AGO.

THE science of medicine started on a grander career from the time of the appearance of the great work of Vesalius on anatomy, published in many editions about the middle of the sixteenth century. This was followed by more correct and minute physiology, and this led to pathology, and so on to more rational therapeutics. Another whole century passed, however, before the art of healing received the aid of that great medical teacher and practitioner, Thomas Sydenham, of England. He was a long way in advance of the thousands of medical contemporaries in England and Europe.

Then appeared Bonetus, with his massive work *Sepulchretum, sive Anatomia Practica*, in 1679, which really laid the foundation of pathological anatomy. Then soon followed Morgagni's *De Sedibus et Causis Morborum*, which confirmed it. Harvey had already appeared and demonstrated the circulation of the blood. Next came Albert Von Haller, one of the greatest, because most learned and wise, physicians up to his time. He developed physiology and all that that signifies. His attention was directed especially to the nervous system, the functions and limitation of the different classes of nerves. He arrived at the conclusion that muscle fibres can be made to contract independent of nerve fibrils. This doctrine long agitated the medical world, and was termed

5

"Hallerian irritability." It is a question with some to the present time.

The great John Hunter was also in the field at that time, and these were followed by Cullen, Brown, and others, the fathers in medicine, to whom we are and shall always be indebted for many fundamental facts in this science and art. Then we arrive at the beginning of the present epoch, the nineteenth century. The improvements of this last period, of course, all are better acquainted with, and they need to be only mentioned by name.

But how much more meagre and crude must have been the instruments and books of the forefathers of old England, when in the fifth century they took their journey westward from old Saxony, and Engle, and Jutland to cross the sea and found a nation in their now beautiful island home. In the ages following, however, the culture of Italy, Spain, France, and Germany overtook them. Finally, in the thirteenth century, Roger Bacon appeared in England and produced his *Opus Majus*, and introduced, or attempted to do so by his teachings, his practical chemistry, which was a great improvement on their wretched chemico-alchemy. However, he was too early, for they put him down and out, preferring old alchemy, which still damaged medicine for many centuries longer. In the lapse of time there was real improvement. It is evident that later on England, with her Oxford, King's College, and Edinburgh University, was the very foremost nation in all the world for improvement in general learning during the last four or five hundred years—excepting the last fifty years, perhaps, when France and Germany have been her peers.

Soon after the year 1700 A.D., as already stated, the art of medicine started on a new career, for the great work of Vesalius on anatomy was published, by which the way was made ready for it. Bonetus had already appeared, and Morgagni, which fairly opened pathology ; and Sydenham was in the field. True, about this time appeared Stahl at the University of Jena, and flourishing in Prussia, established the famous " Expectant " school of medicine, founded upon the pathological *theory* of Van Helmont and the psychological doctrines of Des- cartes. He taught that on the one hand *plethora* and on the other *anæmia* were the principal causes of disease. Also, that " spirits " or vitality, and the *vis medicatrix na- turæ* will bring about health with little or no aid. Then came Hoffman, professor at Halle, who advocated an- other "system" of medicine, basing the cause of disease in a morbid influence from the nervous system, joined with a humoral condition. And a still greater contem- porary was Herman Boerhaave of Leyden, a very learned man, an advocate of the " humoral pathology."

John Hunter arose, and William Cullen, followed by Albert von Haller with William Hunter, followed by Dr. John Brown, the founder of the " Brunonian system," about 1760. His school or system was founded upon what he termed "local excitability." Before him, how- ever, Haller had already taught a general excitability ; that is, that muscle fibres, and perhaps other tissues, may be made to contract by direct stimulation, without the aid of any nerves. This was long resisted by a large part of the profession, and by some even to the present time ; this being mentioned in medical works as the Hal- lerian irritability, but in no way as constituting any sep- arate school or sect.

The present high degree of attainment in the different departments of the healing art is not, then, all born of the present time, as so many think or assume, and not a

few assert. The latter tacitly claim that almost all knowledge in therapeutics has developed, as it were, since yesterday, or certainly in our own day. The fact is, old-
time medicine was wonderfully right in some respects,
and egregiously wrong in others. According to Egyptian
records, as found in their ancient papyrus, from the
times of the Ramases, dating back to B.C. 1400 or 1500,
there were medical *specialists;* or rather, for thorough-
ness, all medicine was so practised by law of the land.

That we may estimate the degree of this wave of increased knowledge and the facilities for carrying on a
more elegant, scientific, and skilful practice of medicine,
let us but look back again from this standpoint into the
very twilight edge of the lingering shadows of the dark
ages, less than two hundred years ago. We shall observe
there the bold admixture of some new truths with old
errors ; and yet so soon as the present time, medical
and surgical improvements are not surpassed by advances in any other department of literature, science,
art, or mechanics, great as they all have been during the
same period. True, some of the old and oldest medical
opinions and principles are our opinions and principles;
and some of their perplexing inquiries are still before us,
as unanswered as ever. We are still learning, and have
much to learn. To this end we must continue to be
teachable and inquiring, though we possess the present
knowledge and skill.

About the year 1700 A.D. we find that the educated
physicians began to think for themselves, and step by
step to call in question their antiquated theories and
practice. The public was ready for it, yes, demanded it.
They had in England already very brilliant contemporaries in other studies and pursuits, such as Sir Isaac Newton, Dean Swift, Samuel Johnson, Pope, Addison and

Steele, Bolingbroke, and Sir Robert Walpole. This was in the reign of Queen Anne and that of George the First, yet it was her Royal Highness that continued the practice of "the touch" for the cure of "king's evil," or scrofula.

We may infer that at this time medical libraries were small and few, and that surgical instruments were fewer still. Especially was this true in regard to those physicians who first came over to these New England shores with our pilgrim and Puritan forefathers between 1620 and 1650.

Boerhaave took the degree of M.D. in 1693, at the University of Harderwick. In 1703 the students at Leyden requested him to teach them chemistry. Soon he publicly contended for mechanical reasoning in Physic, and he became a chief supporter of this idea, though that doctrine sprung from Borelli, a native of Naples, in 1640, who was a professor in the universities of Florence and Pisa. At that date the chemists and the mathematicians divided the whole empire of medicine for some time. But Stahl, in Germany, soon made a great sensation in another direction, for he adopted the idea of "anima," which he conceived to be all-pervading, and of a specific nature, a principle to oppose the physical powers of matter, and possessed of a species of intelligence capable of acting the part of a rational agent, superintending all the corporeal operations. Hahnemann, one hundred years later, accepted this proposition, and adjusted his infinitesimal dose to it. This "anima," or nature, was first stated by Hippocrates two thousand years before.

Boerhaave conceived the existence of two opposite principles in the human body, one of which is constantly tending to corruption and death, the other to health, repair, and life. The first depended on the elementary composition of the organism, the second on the power and energy

of the mind.   Hoffman, a contemporary, rather ascribed
this power to the nervous system, instead of "anima,"
and thus he led the way to a more reasonable system of
physiology.*   Boerhaave endeavored to explain the functions of the body in health, the phenomena of disease,
their causes, symptoms, and even the action of medicines
for their relief, according to the laws of statics and hy-
draulics, and also by the operation of chemistry.

"Boerhaave once intended to prepare a chronological
history of the alchemists, intending thereby to show
that from Geber to Stahl they had all been misled by
one and the same error.   He had taken great pains on
this subject, and the non-completion of his labor is a loss
to this class of history, for he had read over most diligently the works of Paracelsus four times, and those of
Helmont seven times ; his *De Mercurio Experimenta* relates to the transmutation of metals as put forth by Paracelsus, Helmont, Basil, Valentine, and others, a doctrine
which Boerhaave thought should be considered as feasi-
ble.   Impressed with this idea, he tried to consummate
the purification of quicksilver.   With matchless perseverance he tortured it by conquassations, triturations, digestion, and by distillation, either alone or by amalgama-
tion with lead, tin, or gold, repeating his operation to
five hundred and eleven times, and even to eight hundred
and seventy-seven distillations, yet with what result?   It
only appeared rather more bright and liquid, without any
other variation."

"If the mechanical hypothesis embraced by Paracelsus is unable to account for the operations of the animal
system in health, equally defective must it be to explain
the effect of the various diseases to which it is subject.
These views are now exploded, but it will probably amuse

* "Memoirs of Physicians," by Pettigrew, vol. i.

some of my readers, who may not be aware of the extent
to which these opinions were carried, to know that
the same process of reasoning was applied to the remedies proposed for the cure of diseases. In the Philosophical Transactions (vols. 24 and 26) there is actually a table
constructed by Dr. Cockburn, in which are enumerated
the different medicines commonly employed, and in
which the doses are severally adjusted by mathematical
rules and precision, according to the age, the sex, and the
constitution of the patient. The doses are as the squares
of the constitution! Will it be believed that such a doctrine should have been seriously entertained in the eighteenth century? Yet we find a writer (Dr. Balguy) in the
Edinburgh Medical Essays patiently regarding the subject, and attempting to correct what he considered to be
the errors in this table! "You are to dose," says he,
"so much of the medicine as is spent in the stomach and
intestines directly as the constitution, and so much as is
carried into the blood as the square of the constitution,
and the sum into the person's size is the quantity required."

Dr. Brown, in establishing his "system," called the Brunonian system of medicine, divided all diseases into two
great classes; the one he termed "sthenic," and the other
"asthenic," in character; the former to be treated with
lowering remedies, the latter with tonics and stimulants.
Though this "system" after a time died out, these two
generalizing terms *sthenic* and *asthenic* are conveniently
retained, and are still found in some modern medical literature. As Dr. Brown was largely a theorist, while Dr.
William Cullen was a sensible man, as says Dr. Holmes,
the clash between them and their followers was fierce and
long, so as to affect the whole medical world of that day,
especially in England. Finally, the famous Brunonian

"system " was put down, and with it the older systems of former times, all of which had been but the natural outgrowth of imperfect knowledge. With the increasing information in physiology, pathology, and chemistry, those systems soon died out, and a more scientific and rational theory and practice began to prevail.

First Physicians in America.

Dr. Samuel Fuller was the first physician who came to New England. He was one of the first company who came over in the Mayflower and landed on Plymouth Rock, December 22d, 1620. He was one of the deacons of the Rev. John Robinson's church. It is not certain that he was a graduate in medicine, but he is said to have been well qualified in his profession, and proved eminently useful as a surgeon and physician. He not only extended his benevolent labors to the sick among the people at Plymouth, and to the aborigines in the vicinity, but, by the request of Governor Endicott, he twice visited the new settlement of Salem, where he manifested his skill and success in the treatment of the scurvy and other diseases prevailing there at that time ; which was publicly acknowledged by the Governor of the province.

On looking back over the early history of our country, we find that more than a century and a half elapsed after its first settlement before a single institution existed, either for the education of physicians or the regulation of medical practice. Early and ample provision was made for common schools, and even colleges were established ; and there were many learned civilians and accomplished clergymen, but scarcely a regularly educated and scientific physician could be found—none, except here and there one who had come from Europe. Our ancestors had the peculiarities of their native country and the spirit of that age in regard to all things excepting religion, liberty, and morality. Science had been but little cultivated, and it was extremely limited and hypothetical when our ancestors embarked for America. The depressed state of medical science throughout Europe at that time, and their own difficulties and hardships when first here, will account for their disregard of medical education, and its slow progress for many years after.*

The first medical work published in America was written by a learned clergyman of Boston, and entitled "A Brief Guide in the Small-Pox and Measles." It was printed in 1677. Another soon followed by another clergyman, with the title of "A Good Management under the Distemper of the Measles." Indeed, it was the general opinion of the public that ministers ought to acquire a practical knowledge of medicine, as they generally aimed to do, in order to discharge the duties of piety and humanity to the needy people; and though their medical knowledge was limited, their general intelligence and Christian heart aided them to make good use of it, for, unlike the empirics of later times, they were actuated by the purest motives, and for which the people expressed the greatest thankfulness. Yet this state of things did not continue for a very great length of time.

Early in 1638 Harvard College was founded at Cambridge, in New England. Though originally designed to train young men for the evangelical ministry, and to educate the native Indians of the country, as stated by Thomas Sewall, M.D., of Columbia College, it was not long before some of its graduates began to turn their attention to the profession of medicine. Young men, after

* T. Sewall, M.D., of New York.

studying also a suitable time with the most eminent physicians in America, repaired to Europe to finish their education and preparation for practice. Soon this number was
augmented by the graduates of William and Mary College of Virginia and Yale College in Connecticut, the
former of which was founded in 1691 and the latter
in 1700. At a later period several of the graduates of
Princeton College in New Jersey, founded in 1746, and of
the College of Philadelphia, founded in 1754, pursued the
same course. Thus were gradually introduced into this
country a number of well educated physicians, most of
them natives of the country. At this period females were
mostly the accoucheurs of the country ; and any such
thing as a respectable medical library did not exist.
The works of Sydenham, Boerhaave, Huxam, Cowper,
Keill, Douglass, Van Swieten, Mead, Brooks, Heister,
and Lewis were almost the only authors known to the
medical men here at that time, and but few of these in
any one collection. Leonard Hoar, M.D., a distinguished
scholar and physician of Massachusetts, was graduated
first at Harvard College, in 1650. He soon went to England, and having completed his course of medical studies,
received the degree of Doctor of Medicine, at the University of Cambridge in 1653. He was probably the first
native American who graduated in medicine. In 1672
he was elected President of Harvard College, which office
he held till his death in 1675.—*Magnalia*, iv. 129.

A new era now commenced here and in England and
in Europe. The science began to revive. In 1719 the
foundation of the great medical school of Edinburgh
was commenced by the elder Monro, and medical instruction in London was elevated by William and John Hunter, while the University of Leyden was brought to high
repute by Boerhaave and others, and the medical

schools of France began to assume a new character; and our physicians here, while they felt this influence from abroad, perceived the necessity of adopting measures to check the progress of quackery and empiricism, which then threatened to overspread the country.

As early as 1750 the body of a criminal executed for murder was dissected in the city of New York by two of the most eminent physicians of that day, and the blood-vessels were injected, and the whole prepared for preservation in the medical museum, for the instruction of the young men then engaged in the study of medicine. Six years after this—that is, in 1756—a course of lectures on anatomy and surgery was delivered at Newport, R. I., by Dr. W. Hunter, a Scotch physician, educated at Edinburgh. Then, in 1765, a regularly organized medical school was commenced in Philadelphia, Pennsylvania. The second medical school instituted in this country is that of the city of New York, first established under the charter of King's College, in 1767, three years only after that at Philadelphia. But its prospects were soon interrupted by the Revolutionary War, and it was not till the year 1792 that it was recommenced, this time recognized by the trustees of Columbia College, which had been known by the name of King's College before the Revolutionary War. Now this medical school is styled the College of Physicians and Surgeons of New York.

The next great medical school started in this country is the Harvard Medical College, first established at Cambridge, Massachusetts, in 1782, nearly a century and a half after the classical department of Harvard University had been in successful operation. This is now one of the largest and most thorough medical schools of learning in the United States. It was soon after the Revolu-

tionary army had encamped at Cambridge, and Washing-
ton had taken command of it, and military hospitals
opened in that town, that this medical school opened.
The reception of the sick and wounded called together
the physicians from every part of the country; new forms
of disease were developed, and important surgical operations presented which gave rare opportunity for practical
experience.

One of the greatest medical improvements in the eighteenth century was preventing the great mortality from
small-pox by " *inoculation.*"

Sacred to the Memory of *
The Rt. Honorable
LADY MARY WORTLEY MONTAGUE,
Who happily introduced from Turkey
into this Country
the Salutary Art
of Inoculating the Small-pox.

———

Convinced of its efficacy
she first tried it with success
on her own children, and then recommended the practice of it
to her fellow citizens.

———

Thus by her example and advice
we have softened the virulence and escaped the danger of this
malignant disease.

———

To perpetuate the memory of such benevolence,
and to express her gratitude
for the benefit she herself received
from this alleviating art,
this monument is erected by
HENRIETTA INGE,
relict of THEODORE WILLIAM INGE, Esq.,
and daughter of Sir JOHN WROTLESLEY, Bart.,
in the year of our Lord
MDCCLXXXIX.

———

"She died 1762, æt. 73." This epitaph is on a ceno-

* "Life of Mary W. Montague," pp. 305-7.

taph in the cathedral at Litchfield, England. The monument consists of a mural marble, representing a female figure of Beauty weeping over the ashes of her preserver, supposed to be enclosed in the urn, inscribed with her cypher—M. W. M. The following extracts are taken from Lady Montague's correspondence about that time, showing her surroundings.

"*At* BELGRADE VILLAGE.

" As to the balm of Mecca, I will certainly send you some, but it is not so easily got as you suppose, and I cannot, in conscience, advise you to make use of it. I know not how it comes to have such universal applause. All the ladies of my acquaintance at London and in Vienna have begged me to send pots of it to them. I have had a present of a small quantity (which I assure you is very valuable) of the best sort, and with great joy applied it to my face, expecting some wonderful effect to my advantage.

" The next morning the change indeed-was wonderful —my face was swelled to an extraordinary size, and all over as red as my Lady H——'s. It remained in this state three days, during which you may be sure I passed my time very ill. . . . They all make use of it and have the loveliest bloom in the world. For my part, I never intend to make use of it again, but let my complexion take its natural course and decay in its own due time."

"ADRIANOPLE, *April* 1, 1717.

" Apropos of distempers, I am going to tell you a thing that will make you wish yourself here. The small-pox, so fatal and so general among us, is here entirely harmless, by the invention of *ingrafting*, which is the term they give it. There is a set of old women who make it their business to perform the operation every autumn, in the

month of September, when the great heat is abated.
People send to one another to know if any of their family
have a mind to have the small-pox; they make parties
for this purpose, and when they are met (commonly fifteen or sixteen) together the old woman comes with a
nut-shell full of the matter of the best sort of small-pox,
and asks what vein you please to have opened. She immediately rips open that you offer to her with a large
needle, which gives you no more pain than a common
scratch, and puts in as much matter as can lay upon the
head of her needle, and after that binds up the little
wound with a hollow bit of shell, and in this manner
opens four or five veins. The Grecians have commonly
the superstition of opening one in the middle of the forehead, one in each arm, and one in the breast, to mark the
sign of the cross. But this has a very ill effect, all of
these wounds leaving little scars, and is not done by
those that are not superstitious, who choose to have them
in the leg, or in that part of the arm that is concealed.
The children or young patients play together all the rest
of the day, and are in perfect health to the eighth. Then
the fever begins to seize them, and they keep their beds
two days, very seldom three. They have very rarely
above twenty or thirty in their faces, which never mark,
and in eight days' time they are well as before their
illness.

"Every year thousands undergo this operation, and
the French ambassador very pleasantly says, they take
the small-pox here by way of diversion, as they take the
waters in other countries. I intend to try it on my dear
little ones."

———————

Among letters during her husband's embassy to Constantinople is found this from Addison: "Being very

well pleased with the *Spectator* I cannot forbear sending
you one of them and desiring your opinion of the story
in it.   When you have a son I shall be glad to be his
Leontine, as my circumstances will probably be like his.
I have within a twelvemonth lost a place of £2000 per
annum, an estate in the Indies of £14,000, and, what is
more than all the rest, my mistress.   Hear this, and wonder at my philosophy.   I find that they are going to take
away my Irish place too, to which I must add that I
have just resigned my fellowship, and the stocks sink
every day.   If you have any hints or subjects, pray send
me a paper full. . . . Dick Steele and I often remember
you.

> "I am, dear Sir,
> "Yours eternally, etc.,
> "J. ADDISON."

To which Lord Montague answers:
"Notwithstanding your disappointments, I had much
rather be in your circumstances than my own; the
strength of your constitution would make you happier
than all who are not equal to you in that."

The Sultana Hafiten, favorite of the late Emperor Mus-
tapha, invited Lady Montague to visit her.   She went,
and in a letter to an English friend thus describes the
dress of the sultana: "Her donalma was tied at the waist
with two large tassels of smaller pearls" (then a certain
fringe), "and around the arms embroidered with large
diamonds.   Round her neck she wore three chains, which
reached to her knees—one of large pearls, at the bottom
of which hung a fine colored emerald, as big as a turkey's
egg; another consisting of two hundred emeralds closely
joined together, of the most lively green, perfectly

matched, every one as large as a half-crown piece and
as thick as three crown pieces, and another of small emeralds, perfectly round. But her ear-rings eclipsed all the
rest ; they were two diamonds shaped exactly like pears
and as large as a big hazel-nut. Around her kalpac she
had four strings of pearls, the whitest and most perfect
in the world, at least enough to make four necklaces, every
one as large as the Duchess of Marlborough's, and of
the same shape, fastened with two roses, consisting of a
large ruby for the middle stone, and around them twenty
drops of clear diamonds to each. Besides this, her head-
dress was covered with bodkins of emeralds and diamonds. She wore large diamond bracelets, and had five
rings on her fingers—except Mr. Pitt's, the largest diamonds I ever saw in my life.

"She gave me a dinner of fifty dishes of meat. . . . But
the piece of luxury which grieved my eyes was the table-
cloth and napkins, which were all tiffany, and embroidered
with silk and gold in the finest manner in natural flowers,
and towels of same kind, with which I very unwillingly
wiped my hands. They were entirely spoiled before the
meal was over."

The two greatest medical improvements in the last
century were the discovery and use of electrical apparatus
for medical use, and the inoculation of the small-pox to
reduce its great fatality.

The credit of the introduction of inoculation for
small-pox into this country is generally given to Cotton
Mather, who had read in the Philosophical Transactions
of the Royal Society at London, England, that this
method was used in Turkey as a means of protection
against the fatality of small-pox.* During a long time the

* Centennial Address, by S. A. Green, M.D., in 1881, before the Mass.
Med. Soc.

practice had been kept up in Constantinople, where it was brought from farther east, and had met with much success. Rev. Dr. Mather tried to interest the Boston doctors in the subject, but at first it was opposed by them all excepting one, Dr. Zabdiel Boylston, who commenced the practice of it amid the most violent opposition. On the 26th of June, 1721, he inoculated his own son, Thomas, six years old, his negro man, of thirty-six years, and a little negro boy, of two and a half years. They all had the disease lightly, and he was encouraged to continue trying it on others. The colony of Massachusetts had formerly suffered severely from the scourge of the small-pox, and the epidemics of it were periodical. The mortality from it was large and the effect disastrous, so that any help was a boon to the community.

In the course of time inoculation conquered all opposition, and finally became a well established fact in the community. Within the period of one year Dr. Boylston inoculated two hundred and forty-seven persons, and of this number only six died; and during the same period thirty-nine other persons in the neighborhood were inoculated by two other physicians, and all made good recoveries. This low rate of mortality, as compared with that among persons who had taken small-pox in the natural way, was a telling argument in favor of inoculation, which was continued for about seventy-five years, when "vaccination" with kine-pox was discovered by Dr. Jenner in England, which was speedily introduced into this country, to supersede the small-pox inoculation.

In 1799 Dr. Benjamin Waterhouse, of Cambridge, an early Fellow of the Massachusetts Medical Society, published an article entitled, " Something Curious in the Medical Line," which was the first account of kine-pox vaccination that was given to the public in this country.

6

In the year 1800 he published a tract entitled, " A Pros-
pect of Exterminating the Small-pox : Being the History
of the Variola Vaccina, or Kine-pox," etc., and in it he
describes the method he used July 8, 1800, in vaccinat-
ing his son, Daniel Oliver Waterhouse, a lad five years
of age, who had this disease in a very mild way. From
the arm of this boy he vaccinated others.

The Harvard Medical School was removed from Cambridge to Boston in 1810. Since that period it has rapidly grown into a most flourishing institution, near the
great hospitals. The fourth medical school in the United
States is that of Dartmouth College, at Hanover, New
Hampshire, which was established in 1797. In 1798 Dr.
Nathan Smith was appointed its sole professor, who for
twelve years gave lectures on the different branches of
medicine. At this time this school is greatly enlarged,
and has numerous professors. In 1807 the College of
Medicine of Maryland was established in Baltimore. In
1813 the Medical School of Yale College was instituted
under the charter of that seminary of learning, and established at New Haven, Connecticut, in close proximity to
a large hospital. Since which time medical colleges have
sprung up in almost every state of the Union, and a plen-
tiful supply of well educated physicians and surgeons are
provided for the needs of the whole population, and yet
quackery and charlatanism find place to live and thrive.

Although medical colleges are the principal means
by which the science of medicine has been extended
through the country, still there is another class of institutions which has contributed to medical improvement.
We refer to those societies formed for the regulation of
the practice of physic, and the separation and suppression
of quackery. As early as the year 1781, for this purpose,
Legislature incorporated the Massachusetts Medical Soci-

ety—embracing the regularly educated physicians in the state—a body politic, with power to frame a code of by-laws, and regulate the practice of medicine throughout the commonwealth ; to admit suitably qualified members and to expel all disqualified. Similar medical societies have since been incorporated by the Legislatures of all the United States. And these again are organized into the one great American Medical Association, which is composed of delegates from these state organizations, which is supposed to establish a recognized standard of medical qualifications. At the present time, the Massachusetts Medical Society has a by-law that prohibits any one becoming a member of it who advocates either homœopathy, allopathy, or any other ism, pathy, or sectarian practice, but allows female physicians, duly qualified, to be received on equal terms with the Fellows.

Besides these medical colleges and associations for regulating the education and practice of physicians throughout the land, hospitals, infirmaries, and asylums for strangers and the poor, sick, or insane, have been established in all parts of the country. Lately, we have public boards of health, and other supervisions of all things that endanger the health or threaten the safety of the people. Such has been the progress of true medical science in the United States during the past century that it is believed we are not behind any other nation in these qualifications.

We might mention, as some of the new things in the healing art, the superseding inoculation with small-pox by *vaccination* of kine-pox to *prevent* small-pox ; the discovery and use of electro-magnetic machines for medical use ; the stethscope and revival of percussion for physical signs ; a great variety of important surgical instruments too numerous to designate ; the discovery *Acarus*

Scabiei, the cause of the itch ; the acromatic microscope, and all its revelations ; the anæsthetics, as ether, chloroform, and the nitrous oxide gas, etc.; elegant pharmaceutical preparations in great variety ; the ophthalmoscope, laryngoscope, the antiseptics, as carbolic acid, chlorine, iodine, and bromine ; the alkaloids : salacine, morphine ; atrophine, quinine, strychnine ; the calabar bean, ergotine; the sub-cutaneous syringe, and the atomizer, the aspirator, and the stomach-pump; the iodide of potassium, the bromides, and the hydrate of chloral, and dental instruments in great variety and elegance ; artificial teeth, and operations for strabismus, and for diseases in the ear, eye, nose, and fauces.   There is also the fever-thermometer, and the sphigmograph.   But we do not recall all that is new in modern medical science, nor can we describe the
• items, for it would fill a book, so great are the resources of the medical profession of the present day.

If we take our stand in the latter part of the last century and look forward into the beginning of the present, what do we see ? What a contrast with one hundred years before if we take Sir Richard Blackmore, physician to the king in the early years of the eighteenth century, as an example of the rank and file of the profession in his day, and compare him with Sir Charles Bell, as the exponent of the medical profession in the early years of the nineteenth century !

> "Behold ! how brightly he breaks the morning,
> And casts his rays o'er lands unknown."
> *Massienello.*

The first year of the present century brought with it that young medical star whom we all know of as Sir Charles Bell. He then was only twenty-two, and the year before, in 1799, he was admitted a member of the Edin-

burgh College of Surgeons, and then as one of the surgeons of the Royal Infirmary. In 1806 Mr. C. Bell left
Edinburgh for the wider field in London, though the
lecture-rooms there were at that time occupied by anatomists of the very highest repute, among whom were Sir
Ashley Cooper and the famous Abernethy. When he
there joined the Anatomical Society he found himself in
company with Dr. Baillie, Mecline, Abernethy, Cooper,
Sir E. Holmes, Mr. Wilson, and the anatomical professors
of Oxford and Cambridge. He lectured before the medical school in Great Windmill Street, where both the Hun-
ters had been before him. Soon he was elected surgeon
to the Middlesex Hospital, then Professor of Anatomy
and Surgery to the Royal College of Surgeons.

On the accession of William IV. it was proposed by
Government, with the cordial sanction of the sovereign,
to confer the distinction of knighthood on a limited number of the most eminent men in science, and the following
persons were selected : Sir C. Bell, Sir John Herschel,
Sir David Brewster, Sir John Leslie, and Sir James Ivory.
In consequence of his discoveries in the nervous system,
which gave him a European reputation, Sir C. Bell was
called to the London University, now University College.
In 1836 he was invited to return to Edinburgh as a professor in the University there, which he accepted. In
1811 he published his "New Idea of the Anatomy of
the Brain," in which he, for the first time, announced those
views of the nervous system on which so much of his
fame rests.

Sir Charles Bell says : " Human sufferings and human
credulity afford a never failing harvest ; quackery is an evil

"' Which walks uncheck'd, and triumphs in the sun.'

" By much the larger portion of patients received into

the cancer ward of the Middlesex Hospital have pre-
viously spent their last penny, and, what is worse, they
have lost that precious time in which they might have
been cured, in trying the attendance of a set of the
most unfeeling wretches that ever disgraced a country."
In London, after Sir C. Bell had ceased to be a lecturer,
his practice, though extensive, was chiefly in nervous
affections, to which his high reputation entitled him to
the first place.   Consequently he was held in high esteem in professional consultations in all difficult and
obscure cases.

Lord Brougham was a Christian philanthropist, and
interested in the publication of *The Library for the Dif-
fusion of Useful Knowledge*, and this drew out Sir C.
Bell to prepare two papers on " Animal Mechanics,"
which became so deservedly popular, and no doubt led
to the illustrated edition of Paley's " Evidences of Natural Religion," and the Bridgewater Treatises.

His researches in the structure and functions of the
brain, the beginning and the end of the whole nervous
system, to which every sensation is referred and where
every idea originates, led him to announce "that the cerebrum and cerebellum are different in function as in form ;
that the parts of the cerebrum have different functions;
and that the nerves which we trace in the body are not
single nerves of various powers, but bundles of different
nerves, whose filaments are united or bound up together
for the convenience of distribution, but which are distinct in office as they are in their origin in the brain or
spinal cord."

At the present time, all medical men agree that nervous affections in number and variety are on the increase,
relative to population.   All the more so in cities and
centres of culture and competitions.   Time was when

the people generally were content to have things as they had been long before. Not so now. Children are not willing to begin as their parents did, but strive to begin where their parents leave off—in ample homes and surroundings. We now cannot ride fast enough even in the steam-cars, or steam-boats, or with fine horses. Life with us all is too competitive and too fast ; inventions and improvements lead to rivalry, wealth leads to luxury and extravagance and ill health. The pressure begins with too much study in school hours and out, and with this high culture of the intellect the nervous system is rendered more susceptible to all impressions from varying weather and circumstances. Steam-heated houses, carpets, and bad air from furnaces and from much combustion of gas, enfeeble us generally, and so unfit us for endurance.

Hence, nervous affections, consumption, and increasing insanity. But with these come a new and varied class of remedies, especially for the nervous maladies, together with new methods in diagnosticating and treating them, that meets the exigency wonderfully. If we notice the space given to the recent discussions of diseases of the nerves, as in Braithwaite's " Retrospect," where all diseases are treated of in large classes, as those of the respiratory system, those of the digestive system, those of the circulatory system, etc., we find space given to the nervous system is the largest. No doubt the increased demand calls out the sharpness of these researches, for from the study of the diseases of the nervous system we learn more of its nature than from any other source.

In order to bring the subject of nervous affections, as they occurred some one hundred and fifty years ago— then generally called the *Spleen* and the *Vapors*—graphically before the mind of the reader, the very thoughts and

identical words of the old-time doctors must be before
us ; not as one may now endeavor to describe them, but
as the medical practitioners *of those times* themselves
set them forth.   To this end it will not do to quote
either Willis or Sydenham, for they were a generation
in advance of their times, but such educated reputable
medical gentlemen of large practical experience who were
well acquainted with the generally prevailing views of
the profession and the public.   One of the most prominent of those in England was good old Dr. George
Cheyne, the practitioner and philosopher of whom we all
have often heard.   He was one of the earliest considerable
writers on nervous maladies.   Among the rare old medical books in our public library may be found a work by
this " George Cheyne, M.D., F.C.R.Ed.S., F.R.S.," entitled, " Medical, Moral, and Philosophical," dedicated to
the Right Hon. Earl of Huntington, etc., etc., published
in London, England, and dated in preface Bath, August
15, 1730.   Probably this was one of his last works.   He
evidently was an extensive writer, for he also wrote on
the spleen, the vapors, lowness of spirits, the palsy, etc.
He had a large experience, was a devout Christian and
a versatile writer.

George Cheyne, M.D., was born of a wealthy family
in Scotland in 1671.   He was educated professionally
under the eminent Dr. Pitcairne, and graduated at Edinburgh.   He passed his youth in close study and regular habits, but, according to his own account, he afterwards settled in London, and finding the young gentry
and free livers to be the most easy of access and the
most susceptible of friendship, he changed his course of
living, under their influence, till in a few years he grew
excessively fat, short breathed, lethargic, and quite indis-
posed to active exercise.   Having tried the powers of

medicine in vain while still continuing this course of life, he says he finally resolved to adopt a plain, strictly milk and vegetable diet, with a free use of nuts. This, he says, soon began to remove his complaints, so that his size was finally reduced one half. He recovered his strength, cheerfulness, activity, and happiness. He is said to have lived to old age. He wrote an " Essay on Health and Long Life," "An Essay on the True Nature and Due Method of Treating the Gout," "A New Theory of Acute and Slow Continued Fevers," a work on " The Philosophical Principles of Religion, Revealed and Natural." Also a work entitled " The English Malady: a Treatise on Nervous Diseases of all Kinds, in three parts." To quote freely from this last mentioned work will probably give a good view of the theory and practice of old-medicine, especially as it relates to nervous affections in those times ; which we must bear in mind was this side of the year 1720, or about 100 years since the Pilgrim Fathers landed on Plymouth Rock in Massachusetts Bay.

From his standpoint, Dr. Cheyne thought he saw danger to his countrymen from their rapidly increasing wealth and consequent luxury; for he says dissipation was replacing their former diligence, frugality, and simplicity. In warning them of it, he wrote many works and aphorisms. It is said that while some persons were expatiating, in the presence of this cute old doctor, on the goodness and excellence of human nature, he said, " Hoot, hoot, mon ; human nature is a rogue and a moody scoundrel, or why should it perpetually stand in need of laws, physic, and religion ?" In fact, in some respects, Dr. Cheyne appears to have been to the English people in his day somewhat as Benjamin Franklin was to us one hundred years ago. Said he, " Instead of the plain good fare of former times, when the cattle and other animals,—as hogs and sheep, fowl and geese,—were allowed to roam the fields and feed in their natural way, they and we are over fed. The cattle are now doctored and stuffed almost out of their lives, and are made as great epicures as those who feed on them. For by stalling, cramming, bleeding, laming, sweating, and causing them to eat such unnatural food, *nervous disorders* are produced in the animals themselves; and this may be *one* cause of the alarming increase of nervous distempers among us of late. In fact, these causes, pretty nearly as I enumerate them, have been assigned by former physicians and philosophers in former ages and countries to have produced these same effects."

"The Egyptians were the first to cultivate the art of ingenuity and politeness, so they soon arrived to frequent sicknesses, and likewise first brought the medical art to a high degree of perfection. The ancient Greeks, while they lived in their rigid simplicity and virtue, were healthy, strong, and valiant; but in proportion as they advanced in wealth and learning, and became noted for politeness and refinement, they sank into effeminacy, luxury, dissipation, and diseases. They then began to study physic, to remedy the evils which their luxury and laziness brought upon them. Likewise the Romans fell from their greatness and heroic virtue, which had already gained them the world." He thus concluded that the recent rapid increase of nervous diseases had one cause in the increasing luxury and ease of the English people. He quotes Celeus as saying that "these two, luxury and laziness, had first spoiled the habits, health, and constitution of the Greeks, and afterwards those of his own countrymen, the Romans, when they had become possessed of the luxury, as well as the country, of those polished people." "Since our navigation is being so extended we have ransacked the globe to bring together articles of delicacy and variety, and in plenty, for luxury, riot, and excess. Then in large cities, like London, where the nervous distempers are most rife, outrageous, and obstinate, are added a multitude of additional causes peculiar to city life."

However wrong, as well as right, this good-hearted doctor may have concluded in matters of medical theory and practice at that time, yet we shall doubtless find that he, together with Sir Richard Blackmore, physician to the king, and others whom we may quote, will lead us all to rejoice that, so far as medicine is concerned, we did not live before, nor about, two hundred years ago. In the next chapter we will quote Dr. George Cheyne and his contemporaries in their own words, as they relate to the cause and cure of nervous affections in their day.

CHAPTER IV.

OLD-TIME THEORY OF THE NATURE AND CAUSE OF NERVOUS MALADIES.

THIS old-time doctor says: "The most general Sources and Causes of chronical Distempers are—1st. A glewiness, sizyness, or viscidity in the Fluids of the human body, either accidental or acquired by those Persons who were born with sound and good-conditioned Juices; or original and hereditary in those who have brought them so disposed into the world with them from the ill state of health or bad Humors of the Parents, which possibly they may have had transmitted to them from theirs, and so on for many Generations backwards.*

"2d. Some sharpness, grossness, or corrosive quality in the fluids, due to a saline or other irritative mixture thrown into them, or from some crude concretions not sufficiently broken and divided by the digestive powers in the Alimentary tube, retarding the circulation in the small vessels, whereby the stagnant juices become sharp and corrosive, and the salts have time, by stagnation and their innate attractive quality, to crystallize or unite in greater clusters and exert their destructive force on the solids; and this will be still more pernicious and fatal if the Food is not only in too great a quantity for the concoctive Powers to break and divide completely, but likewise if too highly seasoned with Salt, from which the most terrible symptoms will ensue.

* Cheyne on "The English Malady," vol. i., p. 6, 3d ed.

"3d. A too great Laxity or want of Tone, Elasticity, and Force in the Fibres in general or in the Nerves in particular; for a due degree of strength and springyness is required in the fibres to make the Juices circulate and carry on their motions backwards and forwards in a continual Rotation throughout the whole Habit, and to fully subtilize them so that they may easily pass through the finer tubes of the cappillary Vessels, and also through the strainers of the glands, both to throw off those Recrements and grosser parts which are not required for the animal functions and to separate those Juices which are required for the Preservation of the individual.

"The first cause mentioned will obstruct and sometimes burst the small cappillary Vessels, producing Tumours, Swellings, and Ulcers, and will first tumefy and then relax and spoil the numerous Glands, and so stop the secretions and fill the body with vicious and morbid Juices. This swelling of the Glands will cause them to press upon the Nerves, stop or hinder their Vibrations, or Tremors, or whatever else be their Action. The second cause will not only rend, tear, and spoil the vessels, creating acute pains and producing Scorbutick and Cancerous Ulcers and Sores in any or all parts of the body, but will also, by *twitching* the Nerves or their fibres, produce Pains, Convulsions, or Spasms, and all the terrible class of Nervous Distempers.

"The Solids, and the Fibres and Nerves whereof they are woven and complicated, are subject to several Disorders, as from being too dry or too moist, so spoiling their due tone or Elasticity. The first is caused by too rich nourishment, which sends about the circulation too rapidly, producing inflammatory Disorders, high Fevers, and other acute diseases. The second is caused by too great quantity of nutritious or stimulating drinks, more than the Expences of the body requires, soaking and relaxing the Solids of the body and producing slow and cold diseases.

"Another cause of Nervous Disorders is by noxious Particles getting into their Substances, which will gradually alter or spoil their Functions, whatever that may be, whether by Vibrations, intestine-like action and re-action, or however they act or are acted upon to convey and propagate the Sensations of influence of external bodies to the seat of the intelligent Principle; for when the Juices are spoiled and the Blood declines from its due fluidity they become clogged by what I chuse to designate by the general name of Animal Salts. The Nerves and Fibres are thus injured or destroyed for producing their vibrations or tremors.

" Finally, those Diseases are chiefly and properly called Nervous whose symptoms imply that the system of the Nerves and of their Fibres are evidently relaxed or broken. These may be embraced under the following heads:

" Of the General Division of Nervous Distempers.

" I. All Nervous distempers whatsoever, from Yawning and Stretching up to a mortal fit of Apoplexy, seems to me to be but one Disorder, or the several steps or Degrees of it, arising from the want of sufficient force and elasticity in the solids in general, and in the *Nerves* in particular, and in proportion to the irritation and resistance of the Fluids. In treating of Nervous Distempers the disorders of the Solids are chiefly to be regarded.

" II. The most natural and general Division of nervous distempers will therefore be thus: first, into those diseases that, besides their other symptoms, are attended with a partial or total Loss of Sensation, for some time at least. This class will not only comprehend all those cases of Lowness of Spirits, lethargy, dulness, melancholy, and moping, up to the severest pains, or a complete Apoplexy, but also fainting Fits, so common in persons of weak nerves. These seem to be chiefly caused by a grossness or viscidity of the animal juices, which obstruct the vibrations and tremors of the *nerves* from the brain.

"III. Those nervous Disorders which are attended with Loss of voluntary Motion, or perhaps also have the symptom of Shaking, in any or all of the instruments of voluntary motion. These are the Paralytic kind, from a general Palsy, a Hemiplegia (or palsy of half the body), or palsy of a limb, to a deadness, numbness, weakness, or coldness of any part. This class of nervous diseases seems to owe its cause to loss of Tone in the nervous system, or from suspension of nerve vibrations, whereby the Soul is disabled to communicate its Will, or Energy of motion, to the muscular fibres.

"IV. Those Nervous Distempers that are attended with Spasms, Cramps, Convulsions, or violent Pains or contractions of the muscles. Of this kind are all the convulsive sort, from Hypochondriacal and Hysterical Fits, and the Convulsions of the Epileptic kind, down to Yawning and Stretching. These seem to be produced by saline particles, or some noxious steam or acrimonious fluid or wind, that is pent up and irritates the nerves, which provoke violent throws, contractions, cramps, and spasms, until tormenting and wearying out the elastic fibres at last by their strugglings and efforts the destructive matter is discharged or removed. Much in the manner of that struggle which we observe from sulphureous, bituminous, vitrolick, and ferrugineous particles commingling and fermenting in the bowels of the Earth, and acquire such force, violence, and impetuosity as to make houses, Palaces, and Cities shake and tremble, and to overturn Hills and Mountains, and make Rivers, Lakes, and the Sea itself to boil and heave! till they have forced a Breach and rupture.for their passage into the air.   But when the pent-up matter is not vented the solids are overpowered, the patient appears sinking, and after a few languid motions, and life is spent, the contest ends in Death.

"V. There is another common Division or distinction of Nervous Disorders, into *original* and the *acquired*. These differ but little.   It is to be supposed that Mankind were originally made so as not to differ much from a standard of Good Health in their constitutions.   Therefore original nervous disorders have had the same Source and cause with the acquired ones.   It may be a misfortune to be born with tender nerves, but if rightly used and managed, even in the present state of things, it may be the occasion of great Felicity; for it is, or ought to be, a fence and security against the snares of temptation to which the robust and strong are exposed, and into which they seldom fail to run, and thereby reduce themselves to the same, or perhaps a worse state.   I shall observe but two things in regard to these persons.

"The first is, that they should never expect the same strength, nor be able to run into the same indiscretions or excesses of sensual pleasures, with those of coarse and strong fibre, without suffering presently, or on the spot, as the others never can.   No art can make an Eagle of a Wren.   The old proverb is still true, that 'A Venice Glass will last just as long, if well looked after, and even shine more bright, than a more gross and coarse one.' Infinite Goodness will ever bring good out of innocent evil.   We may depend upon it.   I can never be induced

to believe that the omnipotent and infinitely good Creator could, out of choice and election, or by unavoidable
necessity, have brought some into such a state of misery.
No, none but men themselves,—I mean the parents, who
were the instruments or channels of these constitutions,—
could have elected, directly or indirectly, to produce such
effects. It is men and women themselves, by wrong doing, that create the miseries of the human family.

" VI. When the general Causes I have mentioned
came to exist in considerable degree and operate in this
climate, then these Nervous Diseases began to appear
more numerous, and with more atrocious symptoms.
Sydenham, our countryman, is the noted Physician who
has made the most particular and full Observations on
them, and established them into particular Classes and
Tribes, with proper though different Methods of cure
from other chronical and humorous Distempers, though
their nature, Cause, and Cure has been less known than
those of other Diseases ; so that those who could give no
tolerable account of them have called them the Vapours,
Spleen, Flatus, Nervous, Hysterical, and Hypochondriacal Distempers. . . .

" The sensible or compound Fibres, as they occur in
the structure of the body and limbs, are of three kinds.
First, some are loose, soft, and elastic, contracting easily,
being moistened with Blood freely ; and such are all the
muscular Fibres. *Secondly*, others are of a closer and
more compact make, and their Elastic force is greater
and quicker, being moistened with a watery fluid to keep
them from growing rigid ; and such are the Fibres of the
Membranes, Tendons, and Nerves, whose compactness
seems to be the reason of their greater degree of Sensibility. They are evidently endued with sensibility above
those of the first kind, the motion or impression commu-

7

nicated to them being thereby less interrupted, broken, or
lost; and a portion of these, the Nerves, are made use
of to convey the impressions they receive from outward
objects, by the muscular fibres, to the Sensorium in the
Brain, and by it to the Sentient Principle, and then from
it out to the Organs. *Thirdly*, the other set of fibres are
of a hard and rigid make, whose elasticity is like that of
Steel, and not fit for sensation. Of these kind of fibres
are the Bones.

"The *Fibres* in general, then, are much alike, differing
in their composition according to the Uses they were intended for. It is likewise probable that all the fibres of
the body are sensible, more or less, according to their
density, elastic, or distractile quality, or connection with
the Brain, the Nerves being only some of these. To tell
precisely how these latter act is, I am sure, very difficult,
and perhaps impossible; nor do I think it any way necessary to what I have to propose concerning the Nature and
Cure of Nervous Distempers. I am of opinion that these
disorders do not depend on the condition of one kind of
animal Fibres, the Nerves, as is by some supposed, but
that when there is general internal disease then the nerves
suffer, as also in small topical disorders, from external
influences.

"What is the true cause of Elasticity in general, or that
of animal fibres and Nerves in particular, is, I think, an
inexplicable problem, unless we admit of a centrifugal
or 'repelling fluid.' Even the true nature and cause of
Cohesion and continuity itself was uncertain until of late.
. . . . Sir Isaac Newton has shown the analogy of bodies
flying from one another, or a principle of Repulsion to
negative Quantities, whereby he hints a probable reason
of the elasticity and the compression of the air. This
may account for the elasticity or repulsion of all fluids

and solids. For example, that experiment whereby a smooth prism, if rubbed strongly, drives leaf-gold from it, and suspends it until its influence is withdrawn ; also Hawsbee's experiments with an exhausted glass Sphere or Cylinder, when violently turned on an Axis." Here, no doubt, Dr. Cheyne refers to the early manifestation of static electricity by the experiments in his time. He then quotes Sir Isaac Newton as saying that these and other like phenomena may be owing to an infinitely subtle, elastic Fluid or Spirit, existing in all bodies and everywhere.

" The Doctrine of Spirits, to explain the animal functions and their Diseases, has been so readily and universally received from the Days of the Arabian Physicians, and before, down to the present times, that scarce one, except here and there a Heretick of late, has called this Catholic Doctrine in question. Even those who perhaps had courage to doubt it, or examine the matter, to avoid explanation have gone implicitly into the common Dialect, which is still very convenient. This system at first was but rude and imperfect, but having been adopted by Philosophers and Mathematicians, as also by Physicians, they have brought it to a more consistent and less absurd Theory. Borelli gave it great acceptance by receiving it to explain Muscular Motion. Dr. Willis gave it all the advantage of eloquence and metaphor. Bernoulli has added to it a kind of geometry and calculation. And last of all, Mons. des Molieres, in the Memoirs de l'Academie Royal for 1724, has added explanation and comparison to the Natural appearances, and removed some common objections. Dr. Pemberton, on the contrary, has shown its insufficiency in his Preface to Cowper's Book on the Mussels.

" On this it may be stated that the best Eyes or Senses, however assisted, have not been able to discover any Cavity in the Nerves, or in the smallest fillaments into which they are divided. Nor by compressing them by ligatures, stopping the influx, or by stroaking and milching their lengths, are any appearances observed like those in other vessels which we know do carry Fluids in them. True, by tying the trunks of the greater Nerves the muscle it-self will turn Paralytic and motionless, but it will do so by stopping the influx of the Blood; which concludes nothing but this, that nerves are necessary towards the Action of the muscles, whatever be their mode of help.

"It is true, the Newtonian Æther theory advances us one step further into the Nature and relation of things; but here we must necessarily stop, the works of God appearing literally inscrutable to perfection. A few of the first steps we may go in this infinite progression, but in the works of God there is a *Ne plus ultra*. It might be urged that as there is a *Mean* between the least and the greatest, so in substances of all kinds there may be in-termediates between *pure*, immaterial Spirits and gross Matter, and that this intermediate material substance may constitute the Cement between the Soul and Body, and be the instrument or medium of all its actions and functions through its material organs. To conclude, if we must suppose animal Spirits, we may affirm that they cannot be of the Nature of any Fluid we have any notion of; that is, from what we see or know. May not the Brain, in some respects, be like other Glands which certainly do seperate Liquors, seperate a milky liquor to moisten and continue the elasticity of the nerves so that they can play off the Vibrations, Tremors, or Undulations made in them by bodies, or their effluvia, or the Will; and may not these Vibrations be propagated through their Lengths by a most subtle, *Spirituous*, and extremely elastic fluid, which is the Medium of the Intelligent Principle?

" Finally, the *Solids* of the human frame seem to be the proper, the only instrument of Life and animation, while the Fluids were only intended to preserve them in due Plight, Glibness, Warmth, and Tonic Virtue, and to solder and otherwise repair the Wounds, Wastes, and Decay. But it is in and through the Fluids that all Medicines and Medical operations have chief effect."

The effects of gout on the nervous system and the mind were well understood by Sydenham, and others in England and France, for he says, "The body is not the only sufferer, and the dependent condition of the patient is not his worst misfortune. The mind suffers with the body, and which suffers most is hard to say. So much do the mind and reason loose energy as energy is lost by the body, so susceptible and vascillating is the temper, such a trouble is the patient to others as well as to himself, that a fit of the gout is a fit of bad temper. To fear, anxiety, and other passions, the gouty patient is a continual victim : and when the disease departs the mind regains tranquillity." *

The most enlightened nations of antiquity had not made much progress in any of the actual sciences but the mathematical. During the Anglo-Saxon period the general mind of Europe turned from their cultivation to other pursuits more necessary. The Arabian mind, being completely settled in fertile countries and mild climates, enjoyed all the leisure that was wanted for the cultivation of natural knowledge, and it showed acuteness and activity. Besides the rules of Latin poetry and rhetoric, the Anglo-Saxons studied arithmetic and astronomy as practical sciences.† That they attained great skill in calculation the elaborate works of Bede in the 8th century,

* " Works of Sydenham," vol ii., p. 128.
† " The Anglo-Saxons," by Sharon Turner, p. 438.

and time of Alfred and Charlemagne, abundantly testify; while they had but little of experimental knowledge. They had a direct and just mode of thinking, though knowledge was imperfect. Although to teach that thunder and lightning were caused by the collisions of the clouds, and that earthquakes were the effects of winds rushing through the spongy caverns of the earth, were erroneous deductions, yet they were light itself compared with the superstitions which other nations then had and have attached to these phenomena.

Among the disorders which afflicted the early Anglo-Saxons we find scrofula, the gout, or foot-ail, fevers, paralysis, hemiplegia, ague, dysentery, consumption or lung-ail, convulsions, madness, blindness, diseased head, and the head-ache. If we can judge from their numerous charms against specific affections, the catalogue of disorders must be extended. Nations in every age and clime have considered diseases to be the inflictions of evil beings whose power exceeded that of man; then adapting their practice ,to their theory they attacked spells by spells. They opposed charms and exorcisms to what they believed to be the work of demoniacal incantations. The Anglo-Saxons had this same superstition, for their pagan ancestors had the same views and applied these same remedies. Hence we find in their MSS. a great variety of incantations and exorcisms against the disorders that distressed them.

In time, when some of their stronger intellects had attained to discredit these superstitions, and especially after Christianity opened to them a new train of associations, this system of diseases originating from evil spirits, and of their being curable by magical art or phrases, received a fatal blow. It began to decline before they were enlightened by any just medical knowledge, and the consequence was that they had nothing to substitute in the stead of charms but the fancies and the pretended experience of those who arrogated knowledge on the subject. Before men began to take up medicine as a profession, the domestic practice of it would of course fall on females, who, in every stage of society, assume the task of nursing the sick, and of these the aged, as the most experienced, would be preferred.

" But the Anglo-Saxons, as early as the seventeenth century, had men who made the science of medicine a study and who practised it as a profession. It is probable that they owed this valuable improvement to the Christian clergy, who not only introduced books from Rome, but who, in almost every monastery, had one brother who was consulted as the physician of the place. We find physicians frequently mentioned in Bede, and among the letters of Boniface there is one from an Anglo-Saxon, desiring some books *de medcinalibus*. He says they had plenty of such books in England, but that the foreign drawings in them were unknown to his countrymen, and difficult to acquire. We have a splendid instance of the attention they gave to medical knowledge in the treatise described by Wanley, which he states to have been written about the time of Alfred. The first part of it contains eighty-eight remedies against various diseases, the second part sixty-seven more, and in the third part are seventy-six. Venesection was in use, but it was held to be dangerous to bleed when the light of the moon and the tides were on the increase."

" With us ther was a doctur of phisike,
In all this world ne was ther non him like,
To speke of phisike, and of surgerie :
For he was grounded in astronomie.

He kept his patient a ful gret del
In houres by his magike natural.
Wel coude he fortunen the ascendent
Of his images for his patient.
   He knew the cause of every maladie,
Were it cold, or hote, or moist, or drie,
And wher engendered, and of what humor.
He was a veray parfite practisour.

  .   .   .   .

Ful redy hadde he his apothecaries
To send him dragges, and his lettuaries ;
Wel knew he the old Esculapius,
And Dioscorides, and eke Rufus ;
Old Hippocras, Hali, and Gallien ;
Serapion, Rasis, and Avicin.
   Of his diete measurable was he
For it was of no superfluitee,
But of gret nourishing, and digestible.
In sanguin and in perse he clad was alle
Lined with taffata, and with sendalle,
And yet he was esy of dispence ;
He kept that he wan in the pestilence.
Fer gold in phisike is a cordial ;
Therefore he loved gold in special."
Chaucer, in " Canterbury Tales."

The Spleen and the Vapors.

"If the Natives of this Island, either from their peculiar constitution, or the air they breathe, or the immoderate quantity of Flesh Meat they eat, or the Malt
Liquors they drink, or from any secret cause, are more
disposed to Coughs, Catarrh, and Consumption than the
neighboring Nations, they are also no less obnoxious to
Hypochondriacal and Hysterical affections, vulgarly
called the Spleen and Vapours, in a very superior degree.*

* "The Spleen," by Sir Richard Blackmore, Kt., M.D., F.R.C.S., in
London.   Printed by J. Pemberton, at the Buck and Sun, Fleet Street, 1725.

And of all the Chronical Distempers that afflict the Body or disturb the Mind, these two, Consumption and the Spleen, are in this Kingdom the most ripe and prevalent, and either directly by their own power or indirectly by inducing other Diseases, make the greatest havock of the health and destruction of the People. I will now show the Nature, Causes, and Symptoms of the Spleen and Vapours, and set forth the method and medicines which, in my judgment, are the most effectual for the Relief of these afflicted persons.

" If a Phthisis is justly called by Foreigners ' Tabes Angelica,' or the *English* Consumption, because it is most predominant, and in a manner peculiar to this country, I am well assured there is no less reason to give to this Distemper the appellation of the *English Spleen.*

" After Aristotle's ill-immagined System as it relates to medicine, supported neither by Reason nor Experience, had the good fortune to become the medical Philosophy in fashion, the physicians generally gave into the Doctrine of this School, and formed their notions of diseases and their cure in conformity to the Peripatetick principles, which, by a swift growth, acquired great power and Authority; then these errors being admitted into the Æsculapian colleges, and mingling with their conceptions, corrupted the simplicity of the art of Physic. At last a great revolution happened in the Commonwealth of Learning, when the authority of Aristotle, who had held an Empire of vast extent and long duration over the Schools and Colleges, began to decline and go out of credit. It is indeed wonderful that an Hypothesis of Philosophy should continue so long in great repute that had mainly immaginary foundations to rest upon. Now, a generous Principle began to assert its Rights to the Free exercise of *reason*, upon an impartial examination

of things, and to throw off the Yoke of Servitude and
Aristotelian bigotry. As soon after, when the colleges
that were defended by the adherents of this philosopher
had revolted and rebelled against the so-called Prince of
Science, the greatest Part of the heads of this Defection
restored the ' Atomic ' or Corpuscular Doctrine, and then
the practising Physicians turned about with the times,
espoused this hypothesis, and framed their notions accordingly.

"Carolus Piso, a French doctor of much reputation,
endeavored to revise the doctrine of Anaxamenes, one of
the first masters of the Ionic school in Greece, who taught
that Water was the fertile Parent of all compound Bodies; while others of equal Fame ascribed their production to Air, or Earth, or Fire; for each Philosopher had
his favorite Element on which to confer the Honor of
being the sole principle that constituted all the *varieties*
of corporeal Beings. It is true this idea is carried too
far; but if Piso had confined his notions to the vitiated
conditions of the fluids of the human body, as obstructing
the minute vessels by impurities and concretions, and
so causing great distempers, his hypothesis might have
been justifiable."

" It is reasonable to believe that the sad Distempers that
affect the Head, the System of Nerves, and the animal
Spirits, all proceede from depraved serous streams that irritate and provoke the nervous fibres, and drive the spirits
into Disorder and confusion, as will appear before we are
done. All kinds of Fevers take their rise from the Nerves,
since they all make their first insult and impression on
those parts, as appear by the Rigors, Shiverings, and convulsions, twitchings and tremblings, that introduce all
kinds of acute Diseases, whether putrid or inflammatory ;
and when the matter of the Distemper is discharged from

the Nerves and is received into the Blood the nervous
symptoms disappear, and are succeeded by a boiling Heat
in the blood, and usually a great thirst. Now it is certain
that Hypochondriacal and Hysterical patients very often
suffer the same symptoms, and so much resemble these
other diseases that at first they are not easily distinguished.

" I believe that the Serum, which waters the Traces of
the Brain, and passes through the Medulla Spinalis, as
also the minute tracts of the Nerves, is not simple unmixed elemental water, but such as contains the generous and active principle of refined Sulphor, Spirits, and
voltile Salts, separated from the Blood by the ministry of
the glands or the fine pores of the Brain, and then becomes
the true cause of most Maladies. Since some nervous Distempers have gone rather by the name of the Spleen from
the primitive Ages of physick to this time, one would
think some defect of this organ was the actual cause of it.
To settle this question, and find out the office of the
spleen, physicians and anatomists of all ages and in all
Nations have laboured. I myself have opened the side
of a dog and torn off with my fingers the Spleen from its
attachment, and closed up the wound, and the dog lived
well for over a year certainly, and probably much longer.

" Nothing was ever so crudely or ill imagined as the
Hypothesis of the ancients, which has likewise been
adopted by a great part of modern Physicians, concerning the Nature of these diseases called the Spleen and
the Vapours, which I shall show are simply one and
the same thing under different Denominations. As the
primitive Practicioners ascribed Hysterical passions to
noxious fumes and vapours ascending, I know not how,
from the womb, so they fancied that Hypocondriacal affections have their rise from the dark steams and windy

exhalations from the Spleen. Now the fact is, the inte-
gral parts of this organ are not composed of coagulated
Blood, as the Ancients, for want of correct ideas, rashly
affirmed; but the Spleen is a system of membranous
fibres, Nerves, and Blood-vessels, so closely connected that
they leave only little cells and narrow ways, but no Cavity,
to be the receptacle of any recrementicious Liquors or
gasses supposed to be separated by occult Strainers from
the bile or Blood. Some have supposed its office to be
that of another liver, to convert a portion of the chyle into
blood ; others assert that the use of the Spleen is to convey a fomenting spirit or austere juice into the stomach
and quicken the appetite ; while others teach that its austere liquor or leaven is to quicken and enliven the animal Spirits inhabiting the nerves.

" Malphigius, an accurate searcher into Nature, has enumerated these various suppositions and confuted them
by Anatomical observations, yet he acknowledges he cannot tell what the office of the Spleen is. But I shall enter at once upon the Nature of Nervous affections, which
in my judgment arise from the irregular and disturbed
Motions of the Spirits and the irritable Disposition of
the Nerves ; and this was the opinion of Dr. Sydenham
and Dr. Willis, and now I imagine very generally ob-
tains. Upon this hypothesis the anomalous and inordinate symptoms may be accounted for which cannot by
any other way of thinking be effectually unfolded. It
was an odd fancy to consider this bowel the cistern of the
grosser Lees and the sediments of the Blood, and at the
same time to suppose it to be the spring of pleasant
humours and alacrity—*Splen ridere facit*, that is, the
fountain of both Mirth and Melancholy ; at one time to
laugh, at another to cry.

"Nor are the extravagant symptoms of this dire dis-

ease to be accounted for by the Hypothesis of Dr. High-
more and others, who suppose that it is caused by the
crudities and depraved juices in the Stomach, with an
extra distention ; or, on the other hand, a relaxed tone of
the Stomach and defective digestion ; which supposition
the learned Dr. Willis has abundantly confuted. Nor is
Velthusius more successful in his attempt to explain the
Cause of this Distemper in his book, de Firmentatione.

" Symptoms of the Spleen.

" Hypocondriacal men are of a pale, almost livid or
saturnine complexion, or a dark, suspicious, and severe
aspect ; nor unlike this is their temperament and disposition, such persons being very scrupulous, touchy, and
hard to please. Their Pulse is usually slow and weak, and
below the standard of Nature in other men, and often too
swift, like that in a hectick fever. As to their urine, there
is seldom any remarkable appearance that distinguishes it
from that of others in health, except this, that it is vari-
able, thin, pale, and insipid, flowing sometimes in a profuse
quantity in a Fit of the Spleen, like that from a Diabetis ;
or as if increased by drinking Wine or tea in an inordinate
Degree ; or like that of a woman labouring under hysteric passions, called the vapours ; or like other persons
when terrified or fearing danger.

" The very Seeds of this Distemper, like those of Consumption, Lunacy, or Scrofula, are often interwoven in
the constitution, where they lie concealed till some active
ferment unfolds them, and they gradually discharge from
the Blood and exert their force on the nerves; for this
Protius, this Posture-Disease can assume the shape, figure, and part of any and many diseases, and when it de-
velops, the Spleen, which before was of a red, florid colour,

becomes now dark and livid. It puts forth a long train of complaints and a sad variety of sufferings, of which are these : a morbid Stomach and poor digestion, with an eager desire to eat, and some time after with great oppressions and grievous pains, attended with Storms of wind in this recepticle and the parts below, where seem a dark and troubled Region of animal Meteors and exhalations, where opposite steams and rarefied juices contend for Dominion, and so maintain continual War. These ferments and flatulent effluvia, while they infest the cavity of the stomach and Colon, to the great suffering of the patient, strive and struggle for vent, often with great noise, like Vapours and Reeks imprisoned underground in caverns ; hence those belchings and loud eructations, which are the notorious symptoms of this Distemper. Likewise there is windy effluvia assembled in parts adjacent to the stomach and colon, which murmur, croak, and grumble ever, and impatient of confinement, with loud noise roll through the hollow regions in the belly and in the sides, beneath the short ribs, with killing pains from their acrimony and distention. Some of these sufferers cannot sleep, or sleep in great Disorder, while their heads are confused ; no bright thoughts come to them ; they lie dejected and desponding, for there is no more sleep until daylight, and then they wish to keep their bed. . . .

"When we review the symptoms of Hysterical affections, numerous in phase as they are, we shall soon conclude that there is no great difference between them and the Spleen in the men ; though it has been supposed that in females it originates in the disordered Matrix, which was supposed to send up Clouds of Fumes and dark vapours through the bowels into the thorax, heart case, throat, and the brain.

"The chief Symptoms of Hysteria, which appears mostly in women, are noisy Convulsions and Workings in the Intestines, Cholic pains, Nausea, loss of apetite, and depraved digestion, and noxious Ejections of green, sower, bitter, and other Humours; fits of short and difficult Breathing, Palpitation of the Heart, Faintings, Suffocation, Strangling in the Throat, Giddiness, Swimming of the head, dimness of sight, violent Headache, or profuse Laughter, and on a sudden an immoderate eruption of Tears, Inconstancy, Timidity, Irresolution, frequent and sudden change of Temper; in short, all the train of symptoms parallel with those of the Spleen. In this only they seem to differ, that in the Female Sex they are manifested in a higher degree. For in Hysteric Fits, which often approach near the Epileptic, the sufferer is thrown into violent Convulsions, the eyes are distorted, the face is disfigured, the limbs are thrown about in involuntary abandon, the thorax is oppressed, Reason and Perception are suspended, she sometimes falling to the ground, beats her breast, sets her teeth, bites her tongue, and struggles with such extraordinary Force that she is scarcely restrained by all about her. But the true cause of all this in the Fair Sex is partly from the constitution of their Spirits and partly from the delicacy of their nerves.

"This disease, then, called Vapours or hysteric affections in Women and the Spleen in Men, I take to be the same malady, and not different species, and is what neither Sex is pleased to own. A physician cannot ordinarily make his Court worse than by suggesting to such patients the true nature and Name of their disorders. Therefore they are to be considered and treated considerately. . . . . Cholic, Melancholy, and Palsy are diseases that have great relation to Hysteria and Hypocondria. What is commonly called a Hemiplegia, or a Paraplegia, or a

dead Palsy, that follows a fit, seems rather a variety of
Apoplexy than distinct diseases. It is evident that in
great Apoplexy the animal Spirits inhabeting the Brain
are the primary Subjects of that Disease, which, being oppressed, imprisoned, and confined to the Head by the
sudden irruption and assault of some stupefying matter,
can no longer take their Flight from the Brain, nor keep
up their intercourse with the distant organs of motion.
It often happens, the noxious serum that benumbs and
muffles the Spirits is not malignant enough to overwhelm
the brain, only partially; then as a poisonous steam it
passes swiftly through the Paths and Traces there, then,
entering into the orifices of the nerves, stifles with its bad
influence the Spirits that reside in them, whose Motion
being thus intercepted, a Palsy follows in all the Muscles
to which those Nerves used to Send Messages. There
are other palsies, as that produced by the displacing of
some of the lower vertebræ of the back in Children,
when from the distortion of the joint the bone presses
upon the *Medulla Spinalis*, or the Orifice of the Nerves
issuing thence, by which the influx or irradiation of the
Spirits is cut off, and the lower limbs grow feeble, wither,
and shrink away; so that it appears also that the Nerves
have much to do with nutrition. . . .

" In some of the cases of Apoplexy the narcotic, obnoxious shade passes through only one lobe of the brain, so occasions only a partial eclipse, and takes off the motive faculty of one side only, and then the eye of that side is weak-
ened, the face is flaccid, the mouth is drawn to the opposite
side, the tongue is partly benumbed and unable to form
distinct words. In some of these cases, when the mind
directs and intends one word the patient, by an involuntary error, chuses another word, and at once he himself
is conscious that he speaks another thing from that which

he designed, which seems to arise from this : that the traces of the nerves in the tongue being defaced, or closed up, the Spirits that do duty to that organ, though directed to one part, finding those avenues shut up, are compelled to go into those that are not obstructed." Here is an account of a case now known as aphasia, from a lesion of the left fourth lobe of the brain. Aphasia is the mis-calling of words ; aphonia is loss of voice.

"The second sort of palsy, which depends not upon peccant matter exploded and transmitted from the brain, but arises from internal or external causes, thus chiefly affecting the muscle fibres, while the sensitive faculty is not much injured. In some it affects only the legs, or the hand or arm ; in some only the muscles of the neck and hands, whence they shall tremble or shake for many months or years without further decay"—agitans, or shaking palsy.

"Now the universal Rise and Spring of all the internal Causes of this Distemper, the Spleen or Vapours, may be justly supposed to be the faulty disposition or condition of the glandulous strainers in the Stomach, Intestines, Pancreas, and Mesentary ; that is, when the convulsive Disorder begins in the inferior and rises to the upper Parts of the body; and the like Error and defect is in the brain when Hypocondriacal or Hysterical Fits begin there, as they often do, occasioned by noxious Juices admitted into its cells and vacuities, which a regular conformation or function of its minute Pores would have kept back.*

"The reason why men of a Splenetick Temperament excel their neighbors in cogitation and all intellectual endowments is this, that when the Juices, when strained by the glands from the Nerves and Lymphatick canals, and are deposited in the Brain, Stomach, Spleen, Pancreas,

* Sir R. Blackmore on "The Spleen and Vapours," p. 92.

8

or any other Receptacle provided by nature in the structure of the economy, retain a moderate degree of Acidity
and acrimony, they only stimulate, exalt, and expand the
Spirits to such a just degree as enables them to make their
reciprocal Motions with a due velocity, in obedience to
the impulse of the Mind, as well as when employed in voluntary vital Offices ; and likewise to serve as more refined
instruments of the Understanding by a bright and lively
Imagination in all its lighter operations, which by this
means are stocked with a greater plenty of clear, surprising, and beautiful Ideas than are produced in persons of
a different Constitution.

" But if these Juices contained in any bowel degen-
erate and become unduly acid, sharp, pungent, and austere, then they urge and vellicate the Nerves so much,
and irritate and scatter the Spirits in such a violent manner, that the whole intellectual and animal Administration is violated and disturbed, while the Mind is deprived
of proper instruments for its operations, and the Body is
filled with Pains, Spasms, or Convulsive Disorders, as before has been explained.

" The learned Dr. Willis has formed a Theory on this
subject, in part alike and in part different from that I
have laid down, by which he accounts for all these effects by the good or bad disposition of a Leaven or
ferment which he places in the Spleen ; and this he sup-
poses, while it remains in a regular state, is a great assist-
ant and Refiner of the animal Spirits, but when it is
perverted, and becomes too sowre and austere, he makes
it the chief, if not the sole Cause of Hypocondriacal symptoms. But I am so far from confining them to the
Spleen, that I believe all other Organs are in their turn as
much concerned, and some more, as I have before explained. It must also be acknowledged that. in many

cases this Distemper is ingrain and coeval with the Embryo and interwoven and complicated with the very Principles of the life."

" In speaking of medical improvement, I reccon the moderns, who do not entirely follow the old, must in reality be accounted the Fathers, and the Ancients the novices ; for in Two Thousand Years the medical art should be brought to greater ripeness : which, however, I conceive has not been much effected till the Two last centuries. I would pay due regard to the Physicians of the first ages, as historians, or credible recorders of matters of fact, especially of their own Times ; as to what internal Medicines, and what external Applications and manual Operations they made use of ; as also what Opinions of Diseases and what Methods of Cure obtained in their times." . . .

" Great numbers of the ancients being persuaded that the *Milt* or Spleen was not only superfluous but detrimental and mischievous, thought it very desirable that their patients should be freed from such a noxious thing, and applied their Industry through the ages to find out effectual ways and Means to extirpate or Destroy it ; and this they very diligently attempted by internal Medicines, and external Applications, and manual Operations. To prove that they had the custom of cutting out the Spleen from human beings, I cite Seranus Sammonicus, who says, in regard to the Spleen.

" '——*tumidus nocet, or risum addit ineptum,*
Dicitur exfectus faciles auferre cachinnos.'

" In the Kingdom of Persia, so many centuries ago as the Reign of Ahasuerus, the same with Artaxerxes, and Longimanus, the learned Doctor Prideaux makes us believe it was the custom to evisect, or extract by manual operation, the Milt or Spleen from horses and mares that were kept to carry the King's Despatches and Orders to distant provinces of his Empire, to give them more diligence and swifter feet. Messengers employed by the Greeks for like use were called Letter Carriers by succession, or Couriers, who gave up their packets at certain stations to others ready to receive them, and so on to destination. And because they had authority to take from the owners by compulsion what Horses soever they wanted, they were called Astandæ and Angari. The first use of Couriers or Post-horses in Persia is attributed by Xenophon to Cyrus.

" Now the Persians believing that the Spleen was an incumbrance on coursers in their running, they made it a practice to rid these horses of this clog. To prove this I cite the Chaldee Paraphrase upon Esther, chap. viii., v. 10,* the original rendered by the learned Bochart first into Latin, then into English, thus: He sent these letters by the hand of the Couriers running upon Coursers, and riding naked upon Post-horses and Mares, out of which the Milt or Spleen had been taken, and part of the Plant of their Feet; i.e., the Frogs were pared away."

" But there is another authority that this was a custom among the Jews themselves, long before the Babylonish Captivity, and that is from what is said in the Gemara, cited likewise by Bochart, concerning the fifty coursers that (1 Kings i., 5.) Adonijah is said to have provided to run before him; from all these the Spleen was taken out, and the soles or frogs of their feet were pared."

" Nor is it surprising that the Excretion of the Milt should have been thought a beneficial consequence in some respects, for from the same authorities we see that the mutilation of boys (slaves) promotes the Sweetness,

---

* Bochart, lib. ii., c. ii., of his " Hierozoicon " ; Pliny, lib. xi., c. 37.

Strength, and Compass of their voices, and prevents such alterations of them that happens to others at their puberty. Nor is this the only emolument acquired by this excision, for even the faculties of the Mind are after the operation preserved clearer, and seem to be enlarged and improved. For it is well known that the Kings of the East paid the greatest respect to their Eunuchs, not only because they might safely be intrusted with the custody of their Seraglios, but likewise because of their Capability and Wisdom; and therefore, for their penetration and great abilities in the art of government, they were often set at the Head of public affairs as first Ministers of State, and perhaps had the Chief Administration of the Empire, or were the principal Officers and favorites at court. We read of many such cases in profane and sacred history."

"Turnebus says:* 'The Ancients evidently believed the Spleen was rather noxious than useful, and so for the most part it is ; especially when it swells, and sometimes seems willing to possess the place of other Bowels, and to invade the Region of the Heart.' 'For this inconvenience, and others more grievous, they attempted to cut it away, either by Medicaments called Wasters, or Consumers of the Spleen, so by Degrees to destroy it, or to burn it out by a red-hot iron ; or lastly, to take it out wholly from the Body by excision.' Of this class of Philosophers and Physicians was the celebrated Erasistratus, a near descendant of Aristotle by his daughter, and contemporary with Theophrastus, who were promoters of this doctrine.

"P. Ægineta relates the precise manner of performing this Operation.† That is, 'let the Skin incumbent on the Spleen, already raised by a Hook, be quite burnt

* Turnebus, lib. vii., p. 15, cited by Martinius in voce Lien.
† P. Ægineta, lib. iii., c. 18.

through by an oblong red-hot iron, so that two crusts or Escars may be made by one impression only, and in the same wound they burnt the Spleen underneath.' In the same chapter he says that Marcellus did indeed use a different way of burning the Spleen in human Bodies, and tells how their two methods differed. By which it appears that this Inustion, or burning of the Spleen, was practised by more than one method in his Time, or that different operators proceeded by a different manner. Ætius, who lived near the time of P. Ægineta, after enumerating a long list of inward medicines, says : * ' If these all prove fruitless, then recourse must be had to the Protection of Burning ; ' and after prescribing the preparatory Bleeding, and evacuants by purging, he adds, ' If the whole body be not first prepared for the operation, malignant ulcers have often, or commonly, happened from the crusts or Escars.' We see also that Trallianus alludes to the same when he says,† ' You know that I cured a Soldier, who, when he came to me, had a sore in the region of the Spleen, burnt out with a red-hot Iron, whom I purged for two or three days with medicines by the mouth, and then used great care in prescribing all proper Remedies.' On the contrary, Absurtus and Cœlius Aurelianus declare their disbelief of this old practice ; that is, they say the animals and men could not survive the operation. That is their opinion. Many eminent Surgeons did not believe it possible to restore an intestine lapsed into the Scrotum by Incision, nor that it is possible to cut for the stone above the Os Pubis, as now performed, and is allowed to be as practicable as they are useful.

"Again, those Ancient Physicians who did not try to

---

* Ætius, tet. iii.   † Trallianus, lib. viii.

cure the Spleen by Excision or Inustion, *endeavored* by other external and internal remedies to waste and destroy its very substance. I request the reader to observe here that I do not state or think that these supposed powerful Medicines given for this purpose were ever really endowed with such actual virtue. But if I prove they believed the Spleen might and ought to be consumed away it is sufficient to prove this old doctrine. . . .

" Cœlius Rhodigenus asserts that near Cortyna in Crete the cattle have no apparent Spleen, which the Physicians inquired into most strictly, and found a plant growing there, by feeding upon which the Cattle diminished their Spleen ; when they invented a Medicine for the cure of their Splenitick patients. This plant is the Scolopendria or Asplenium. Dioscorides affirms that this plant wears away and finally destroys the Spleen. This author, who, according to Suidas, lived with Cleopatra in Mark Antony's Time, speaking of this herb, says : ' The leaves have a Virtue, if boiled in vinegar and drank for forty days, to consume the Spleen.' He likewise made a cataplasm of these leaves, to be applied to the region of the Spleen. Pliny says the same thing, so does Cœlius of Asplenium. Galen ascribes the same power to Spleenwort, or miltwaste, in his sixth book de Simple Medica. Celeus repeats the same, with Dioscorides. Myrepsus calls Asplenium or Scolopendria, *Milt-waste*, the same as our English word Spleenwort and Fingerfern. Myrepsus also prescribed his Spleen-cutting Antidote, which he called Antidotus Splenotorum, having no less than sixty different ingredients, consisting of all the enemies of the spleen that he ever heard of."

" About the Eighth Century a great revolution took place in the Commonwealth of Learning; for the last blow being given to the Roman Greatness and dominion

in the reighn of Augustulus, after it had been long weak-
ened by the Revolt of the Provinces in the East and
the violent irruptions of northern Nations into Italy
and the Provinces of the west,* Science and polite Literature forsook Greece and Rome, for Empire and
Learning rise and fall together, and made their Court
to the new-comers and conquerors, the Saracenes and
Arabs. In the mean time a dark Night of barbaric
Ignorance, introduced by barbarous Arms, overcast all
the Nations of Europe, while the schools of Philosophy
were laid waste and Desolate ; nor was the voice of instruction heard in the cities: then is it a wonder that
the Colleges of Physic should be likewise shut up and
abandoned? Whence, for the space of Four or Five of
those Dark Centuries, this art lay rude and uncultivated,
except what was acquired anew from the Arabians,
who were little more than Commentators on Hippocrates
and the Greeks, till Learning began again to break
through those clouds and throw off the almost total
eclipse in which it had been involved for so many
Hundred Years. The first restorers of learning were
Frifingensis and Reuclin in Germany ; Angelus Politi-
anus, Hermolaus, Barbarus, Poggius Florentinus, and
the celebrated Laurentius de Medici, the great patron
of learning and learned men, and the Father of Pope
Leo the Tenth, who himself was a great promoter of
letters, and many others who at great trouble and expense procured Copies of the Ancients, and by incredible perseverance dispelled the Night of Nations. These
worthies called back from a long exile the Arts and
Sciences into Europe."

"It is true, however, that the Art of Medicine was
not so entirely extinguished, but that about the Year

* " The Spleen," by Sir R. Blackmore, M.D.; chap. iv., p. 144.

Eleven to Twelve Hundred two of the later Greek Physicians, Actuarious and Nicholaus Myrepsus, flourished, and in some degree kept up the repute of Physic, at least preserved it from utter extinction. But after the art of medicine had been at a full stop, after the fall of the Roman monarchy, its growth having been checked and stinted by some unaccountable and fatal incidents until the Sixteenth Century and some after, when it and Philosophy revived, and Physicians abounded in all Nations and Countries. But though Physicians were multiplied and Writers were endless, yet, 'till the last century, the noble Art received but little true increase or Advancement.

"About those times the Enquirers into Nature, discerning the vanity of the System of Aristotal, entered upon a rational way of investigation, and laying by that old empty Scheme, applied themselves with diligence to experimental Philosophy. The medical men who always fell into the ranks of the *prevailing* notions of the Philosophers of the times deserted now the unprofitable *Sect* of Methodists, who relied upon elaborate Compounds of plants and drugs more than on the knowledge of disease and Nature. Instead of improving the healing Art by repeated experiments, and finding out the real value and Virtue of simple medicines, vainly imagined to advance it by an immense luxury of elegant formulæ, and huddled mixtures of many ingredients. That is, they tried the Empirical way of Hippocrates and the physicians of the long line from the earlier ages. So, instead of going forward they were always going backward, and retarding the progress of medicine until the last century or so, since which it has advanced more than for Two Thousand Years before."

CHAPTER V.

THE TREATMENT OF NERVOUS DISTEMPERS AS THEN PRACTISED.

"NOW the question comes, How shall we cure nervous disorders? From what has been explained in the former parts of this work concerning the Sources and Causes of Chronical and Nervous Disorders, there will arise Three Indications (in the Cure of all Acute and all chronical Nervous Distempers) from the three principal Causes, concurring towards their production.*

" *The First Intention* in treatment is that which has the greatest influence on all the rest, and will often, in the milder cases, render the other two less needed, or sometimes even unnecessary. The *first intention*, I say, will be to thin, dilute, and sweten the whole mass of the fluids, to destroy their Viscidity or glewyness, to open the Obstructions thereby generated, to make the Circulation full and free, and the Perspiration and other Secretions flow in their due proportions. This, if fully obtained, lays the foundation of all the rest of the Cure, and will even, during that time, take off the violence of the symptoms and make their intervals greater.

" *The Second Intention* will be to divide, break, and dissolve the saline, acrid, and hard Concretions generated in the small vessels, and to destroy all sharpness and Acrimony lodged in the Habit, and so make the juices soft, sweet, and balsamick. This will be more readily effected if the *first intention* has been successfully persued, for it

* " Nervous Diseases," by G. Cheyne, lib. 3, part ii., p. 112, sixth edition.

is the obstructions in the small vessels that stagnate the juices and corrupt and putrefy the fluids, and by giving time and occasion for the smaller saline particles to accumulate and exert their innate quality of attraction and chrystallization, and so unite in great clusters or concretions; so when the Blood is made thin and fluid, these particles are kept dissolved and further apart until they are thrown out of the system.

" *The Third Intention* in order is to restore the Tone and elastick force, to crisp, wind up, and contract the Fibres of the whole System, which I fear is the most difficult to *complete*. If this could be always and totally effected the Cure would be a true Rejuvenation, and no one need grow Old or die. However, there are not wanting means to effect this *intention*, at least in some degree, if judiciously chosen for suitable persons.

" These are the Three General Intentions to be persued in the treatment of, and towards the total and perfect Cure of, Fevers or Nervous Disorders of whatever Sort or Kind; nor are they ever to be confounded or blended, at least not in the first regular attempts towards a cure. For in the first intention we aim to dilute and clear the juices of the body, in the second we give active and ponderous medicines to correct the tissues, and in the third we use tonics and astringents, which would unavoidably interfere with and counteract each other. Therefore these *Three intentions* must be religiously followed out separately, and in due order.

" What the time necessary for *each* Intention must be it is impossible to determine and lay down, for that must be according to the violence of symptoms and the obstinacy of the Distemper. But to give some general idea or notion of the time, that which is required for the *first* may be conjectured from the state of the Blood. If the

Size on tne top of it is much gone, and if the color is,
and it is easily yielding to any dividing instrument, and
the proportion of the Serum to the globular part upon
bleeding (a few ounces only being taken for the trial) be
such as they are commonly found in well persons, and if
the Serum be clear or not too tawny, then may it be concluded that the *first Intention* has been persued sufficiently.   The *second Intention* may also be guessed from
the healing up and cicatrizing of any ulcer or sores, or
the Cure of any acute Pains or Paroxisms.   The *third* is
obvious after these two are ascertained, by the strength
and vigour of the body and the vivacity of the Spirits, and
the regularity and ease of all the functions which necessarily follow upon the third and last ' *intention* ' being followed for a due time and in a proper manner.   But that
every practitioner may more certainly judge of prosecut-
ing each of the several *intentions*, I will now give an account of the different changes that are brought about in
the treatment of Chronical Nervous affections.

" *The blood*, as it flows in the larger vessels by the orde-
nary course of the circulation, seems to be a uniform
mass, much like cow-milk, but when drawn out of these
Vessels, and left quiet without heat or motion to settle
in the open air, it seperates into two Parts, one of a more
glutinous and solid texture, called the Globular, and the
other of a more thin and fluid nature, called the Serous
part ; and both these are found in different proportions
and of different natures, consistance, and color, according to the Disease of the persons in whom they are found.
I shall only here mention *Three* of these different states,
wherein the distinguishing Marks are most evident,
though there are many intermediate Degrees between
these ; but these will include them all of every shade.

" The *first* is, when the globular part is of a moderate

cohesion and firmness in a pretty equal proportion to its Serum and of a red and scarlet color when exposed a due time to the air, and the Serum is about the consistance of common water, pretty clear, and almost insipid, or, at least, not biting saltish. This I take to be pretty near the State of the soundest and best Blood. Secondly, the second state of the Blood is when the globular and groumous part is in a far greater proportion than the Serum, more thick and viscid, having a glew or size on its top, of a blueish at first, then afterwards of a whitish or tallow color, increasing sometimes to half or more the thickness of the whole, the Serum being in a smaller quantity, and of a yellowish or tawny cast, sharp, acrid, and saltish taste. This seems to be of a middling nature (I speak not here of that accidental size generated by the Nitre of the air, as in catching a Cold, which evanishes in a few days by proper management), between the best and the worst, and is common in Pleurisies, Rheumatisms, etc. Thirdly, the last state of the Blood I shall here speak of is where the fibrous or globular part is scarce any at all, while the Serum is above ten or more times the quantity of it; when the globular part swims like an Island amidst the ocean, the serum being sharp, salt-ish, and urinous to the highest degree in its taste. This I take to be the worst state of the Blood, like those in confirmed Consumption or Dropsy, and other mortal Distempers. But in all these Three states of the Blood the Sharpness, Heat, and Acrimony may arise to almost an equal degree, even to that of the worst state, of which we have no means of judging but by the Taste, which is gross and inaccurate, and therefore we must be content with this probebility.

"The first of these is commonly called good Blood; the second, rich Blood; the third, poor Blood. But to apply

this more particularly to the Diseases I am now treating
of : In all Nervous disorders produced by excesses, especially after the meridian of life, the Blood is generally sizy
and viscid, like that of the second state just described.
I have not for these many years let Blood of any one (but
an ounce or two to make observation on, of which I have
had innumerable instances) who being subject to Nervous
Distempers of some sort, as Lowness, Vapours, or Melancholy, and have not had it sizy, rheumatic, and viscid,
with a sharp-tasting, yellow serum, in some more or less
degree.   I have always observed the Blood of the youth-
ful, and those under the meridian of life who were subject to violent Nervous disorders, to be hot, acrid, and
sharp-tasted, though the colour and consistance might be
tolerably good, but found it was occasioned by sipping
too much of wines or distilled liquors.   But if the viscidity of the juices was produced by over proportion of
food received into the Habit, the weakness of the solids
and slownes of the fluids being consequent, obstructions
must necessarily follow in the vessels and glands, in the
liver, mesentery, etc., and then stagnant juices putrefy,
corrupt, turn acrid and corrosive, then crowding on and
through, tear and injure the solids, from whence arise the
highest Pains of Nervous Disorders.

" *The Method and Medicine proper for the First Intention.*

   " In order to attenuate the Juices, to break the cohe-
sions in the parts, and to destroy their viscidity, so as to
let it flow in the smaller vessels with ease, those Medicines are to be chosen chiefly which either by their nature are the most active, or by the shape of their particles
are the most dividing and insinuating, or by their weight
are endued with the greatest force and Momentum, or

which from experience are found to be most effectual for producing these ends.

"I must mention here the necessity, before any Course of treatment be entered upon, in all acute cases, of premising the common and proper general evacuations, as Bleeding, Purging, Vomiting, etc., some one or more of which will always be found necessary to lessen the quantity of corrupt Blood and to clear the alimentary tubes. Among the chief and principal Medicines are Mercury and its preperations ; Calomel, Precipitat *per se*, Quicksilver, Silver-water, Æithops Mineral, Cinnabar of Antimony, Bezoar Mineral, Crude Antimony, Bezoardicum Joviale, Salts of Tin, Ens Veneris, and the like. Next to these are the woods: Guajacum, Sassafrass, Sarsparilla, Aloes Lignum, and Nephriticum. In the third order are the fixed Salts, as Tartar, Sal Wormwood, Broom, Fern, etc.

"In giving Calomel, however it may be managed in cases of other nature but here, for the cure of Nervous troubles, where is supposed weak solids and tender bowels, it always must be necessary to give it in the very smallest doses, as an *alterative* only, and not as an evacuant. For example, in one, two, or three-grain doses, once or twice a day, and thus it will insinuate into the smallest capillaries and do good.

"The medicines next to calomel to be used in the *first intention* are Æithops Mineral, Antimonietum, Cinnabar, but especially the Mercurius Alcalifatus. That which I generally prefer for viscid juices in persons subject to Nervous disorders, in very low and bad cases, is Cinnabar of Antimony. The more robust may bear the Æithops. . . . . But for the young or the delicate Cinnabar is the remedy, for it can be used for a long time. Its effecacy in the Epileptick, and in convulsions in children, and in-

deed in all chronical and nervous distempers, is widely
known and acknowledged.

"There is nothing I could more earnestly wish were
brought into the common practice of Physic than the
more frequent and general, but cautious, use of the preperations of Mercury and Antimony in these chronical and
obstinate cases, especially when given with only a thin,
cool, and mild diet, so as to answer in this first intention
towards a total cure.   Dr. Charlton, who had the licenc-
ing of the Quacks in King Charles the Second's Time,
told on his Death-bed that all the real successful Cures
performed by the Montebanks of his time were owing to
the preperations of Mercury and Antimony *only*.   The
Mercurius Alcalifatus (Quicksilver and Crab's Eyes,
prepared and rubbed together until the first disappears)
is an admirable Medicine.   Then, when we see Mercury
boiled in plain Water only, without loosing much estima-
tible Weight, if this water be taken in small doses and
for a long time, will produce such sensible effects as I
am convinced it will, we may easily conceive the value
of any of its preperations.   I would only say, add the
use of the wild Valerian, so much commended by Fa-
bius Columna, in all Nervous cases, especially in the
worst.   The tea or powder can be given either with
Black Hellebore or the Mistletoe.

"*The Medicines proper for the Second Intention.*

"When the *first intention* has been successfully pursued, so that the Blood is restored to its due degree of
Fluidity and mildness, when the Acuteness of the Pains
and the violence of the Symptoms are lessened by those
ponderous Remedies, and the Fits and the Paroxisms are
less severe or frequent, then the Medicines of this Class

may be employed. Those medicines are to be selected
here that are of the most active and volatile kind, which
have a penetrating and searching steam or vapour flowing
out of them, that can most readily pervade the solids and
get into the inmost recesses of the Habit, for such seem
to be the most effectual for this stage of the treatment.
The fœtid and volatile substances are the chief, and those
which emit the strongest effluvia, as the volatile Gums
and Juices, the volatile salts and spirits, the phosphoric,
and for external use the Saponaceous substances.

"The principal Medicines of this Tribe are Gum Ammoniacum, Galbanum, Assafœtida, Sagapenum, Myrrh, Guajacum, Camphire, Castor, Amber, Salts of Hartshorn, Salt
and Spirits of Human Sculls, Garlick, Hors-radish, and
the like. But Assafœtida is the first in importance ; the
pure and unadulterated gum is known, for it shows white
when first cut through its mass, then on exposure to the
air turns red or pinkish, which accords to the opinion of
the Ancients, by whom its Virtues are celebrated with
praises ; * and for quieting, it certainly still deserves the
greatest praise. In those complicated cases of Asthmatick
and other Pulmonic affections, the Ammoniacum, Galbanum, and Sagapenum are very effectual.

"Medicines proper for the Third Intention.

" When the two former *Intentions* have been persued
for a due time with satisfactory success, so that the ur-
gent symptoms are abated and tolerable ease is obtained, it will then do to enter upon the *third intention ;*
which should bring more comfort, courage, and Spirits to
go through with it to a complete Cure. For the medicines here are more grateful to Nature, strengthening

* Vide Plinii Hist. Nat., lib. xix., cap. 3.

the digestive organs and making all the functions more full and strong, so that Vigour and Cheerfulness flow in daily. To perceive this is a most agreeable entertainment both to Physician and Patient. This pleasure I have often enjoyed.

" The Medicines suitable for this *Intention* are those of the Strengthening kind, which contract, corrugate, wind up, and give firmness and force to the former weak and relaxed solids, fibres, and Nerves. Of this Tribe are the Bitters, Aromaticks, and Chalybiates, such as Jesuit's Bark, Steel, Gentian, Zedoary, Calamas aromaticus, Snakeweed, Camomile flowers, Wormwood, Oak Bark Acorns, and the mistletoe; also the mineral and vegetable sub-acids each and all have their proper place. The most wonderful strengthener of the solids in a great variety of Nervous disorders is the Jesuit's Bark, even in those frequent cases laboring under internal humour of the Gout, for it seems to drive it to the tip of the lower extremity, and so out, especially if we follow it with the chalybitate waters.

" I come now to the *Dietetic* management, that part which has the greatest influence in the Cure of difficult Nervous distempers, and without which the best of prescriptions fail to succeed. A strict diet, with much exercise, or even a total Milk diet, seldom fails.

" There is a more transient species of Vapours, which very commonly seizes young and temperate persons, otherwise strong and healthy, of pretty sound Juices and firm Solids, which affects with Disgust of everything that used to amuse or please them ; a certain Tediousness of Life, a Lowness of Spirits, with languor, Restlessness, Heaviness, or Anxiety, and an Aversion to Exercise either of the mind or body, and sometimes with violent headaches, or dimness of sight ; which symptoms, as they will come on without apparent Cause, so will they go off as unaccountably in some short time, with or without medicine or means used for their Cure. The way of treating such transitory symptoms is by giving Nature a Fillip, to quicken the circulation, as by eating at the next meal some savory and relishing Delicacy and drinking some generous wines or liquors. The fact of this Cure has been too often tried and repeated with success to admit of a doubt. It is to be feared that this has been the cause of Advice to others, those of weak Nerves and low Spirits, to drink a Bottle heartily *every day*, or take drams, or a bowl of punch, and to use salt Sturgeon, red herrings, Anchovies, pickled Oysters, Salmongundy, Ham, and other Pickles and potted foods, as a Provokative (apetizer). But I caution People not to try this Cure frequently, as in certain cases when these symptoms grow stronger, more frequent, and deeper rooted; then it is worse than no remedy, for it will increase the disease.

"The name of Distempers of Patients we must regard as Sacred. *Res sacra miser*, and Nervous distempers especially seem to be under some kind of Disgrace or imputation in the opinion of the vulgar and unlearned; they pass among the multitude for a lower Degree of Lunacy, and the first Step towards a distempered Brain. The mildest construction they can conceive of them is Whim, Ill Humour, Peevishness, or extreme Particularity. In the Sex, it is Daintiness, Fantasticalness, or Coquetry. So that often when I have been consulted in a Case, before I was acquainted with the character or Temper of the person or her friend, and found it to be what is commonly known as Nervous, I have been in the utmost Difficulty when desired to state or define or name the Distemper, for fear of affronting them, or seeming to fix a reproach on a family or Person. If I called the

Case glandular with Nervous symptoms, they sometimes concluded, apparently, that I thought them Pox'd or had the King's Evil.   If I said it was Vapours, Hysteric, or Hypochondriacal Disorders, they thought I called them Mad or Fantastical ; and if they were some of those who value themselves, and fearing neither God nor the Devil, I was in Hazard of a Drubbing for seeming to impeach their Courage, and was thought as rude as if I had given them the lie ; and in some cases, from this too frank opinion, I have lost the care of them.   Notwithstanding all this, the Distemper is a bodily Disease, as much as the Small Pox, or a Fever.   The truth is, it seldom, and I think never, happens only to those of the liveliest and quickest natural Faculties, who are the brightest and most gifted, so equally are good and bad things of this life distributed.   For I seldom ever observed a heavy, dull, clod-pated Clown much troubled with Nervous disorders, or at least not to any eminent Degree.

"In the *General Treatment* of nervous Diseases we may prescribe a quarter of a Pint of Viper Wine—a half gill in the morning and a half gill at night, after an electuary, or the night pills without steel, which is usually very proper.*   Give also, perhaps, tincture of castor, spirits of Hartshorn, Tincture of Assafœtida, or Compound Spirits of Lavender, or sal ammoniac succinated, taken, mixed together, to thirty drops, or seperately, either of them in same dose, in any fluid ; for it is certain here is required the most generous, active, and penetrating or pungent remedies, both to open the obstructed Nerve Passages and to enliven and rouse up by their instigation the oppressed Spirits, and so enable them, the Spirits, to irradiate the muscles and organs, and make their reciprocal flights in regular order and with due celerity.   When

* "The Spleen," by R. Blackmore, chap. ii., p. 201.

opiates are given judiciously, in respect to time and
condition, for extreme pain or invincible wakefulness, it
is a blessed medicine, notwithstanding many Gentlemen
of the faculty oppose it, because Opium will tye up
the noxious fumes and matter from the Spleen or matrix in the Nerves and fix the Humours in the Blood, dis-
tract and confound the Brain, and make the whole body
dull and the Head muddy. But this is not necessarily
so, for it is neutralized by the pain or noxious matters."

" In treating nervous disorders after the foregoing plan :
first are those we may suppose have weak solids. Give
in the first intention, vomits, and follow with calomel in
very small doses, as one, two, or three grains, once,
twice, or three times a day, as an alterative. Then begin
with the chief remedies, as iron, or *mercurius alcalifatus*,
or precipitat per se, quicksilver, silver water, mithredate,
diascordium, ætheops mineral, cinebar, antimony nigra,
bezoar orientale and lunar, bezoardicum joviale, salt of
tin, *cus veneris*, mistletoe, castor, valerian, musk, salts
of amber, assa-foetida, guajacum, myrrh, sagaphenum,
ammoniacum, *sanguis deaconis*, and other choice gums.

" The more popular remedies are : laudanum, opium,
salts of wormwood, garlic, soot, sulphor, phosphorus,
vitrol of mars, spirits of human skulls, essence of vipers in
infinitesimal doses, zodoary, maidenhair lozenges (to be
dissolved slowly in the mouth and swallowed for hypochondriasis), prepared crabs' eyes; also coral with pearls
in equal parts, half scruple per day; burnt hart's horn
shavings, hira picra, elixer salutis, or tinct. sacra, taken
daily for a year. But in all cases where there is pallor and
debility give the finest filings of iron and steel ground
more fine with white sugar candy, one scruple per day
for the month at a time.

" Excepting all rheumatics, assafoetida quilted in a linnen fold and applied to the sides of the thighs of females and bound on with flannel bandage can do more good for the vapours than any other remedy, for it draws downward the spirits from the upper organs and gives them rest equal to opium without leaving depression behind it. Opium is always to be avoided as much as possible in all this class of disorders. But suitable *diet* and appropriate *exercise* in the open air, when it can be, are vastly important to be enjoyed, *in every case*, without exception, only when utterly impossible."

"Let us not forget," says Dr. Cheyne, " the dietetic management, that part which has the greatest influence, especially in the cure of nervous and chronic distempers, and without which the best and truest remedies fail of their effect. If we make inquiry into the practice of the early and purest ages of physic and notice the greater and more universally approved writers in the healing art, we shall find that diet was considered no such contemplable help towards the prevention or cure of diseases as is now taught and practised. On the contrary, we shall find the works of all the more successful practicioners full of particular directions and advice on this topic in every disease they treat of. And what is, and will be ever, admired of the ancients is their *method of cure*, their soundness of *rules* and maxims, and the solidity of their intentions in following the indications of nature. Hippocrates, the father of physicians, thought a strict regimen of diet of such consequence, both to the well and the sick, though in different degrees, that of about ninety books of his which remain, or that pass under his name, there are eight of them which treat of this matter mainly ; and through all the rest of his works he mentions much more of his *dietetic management* than of the assistance he obtained from medicines."

" It will be always true, so long as we have such bodies, and having to earn our bread by the sweat of the brow, that only temperance, and by times abstinence, air, exercise, diet, and propper evacuations can preserve health, life, and gaity, or cure chronical diseases, I mean generally, and the contrary will always destroy them, for they ever mutually expel one another, like fire and water. Even Homer, three thousand years ago, could observe that the *Homolgians* (the Pythagoreans, who were milk, cerial, and vegitable eaters) were the longest lived and most endurable of men. It is observable that Hippocrates, Galen, Celsus, and others of the principal fore-fathers of physic, cured by means of diet, evacuants, air, and exercise mostly, even as well as we do with all our knowledge of animal economy, *materia medica*, mathematics, natural philosophy, chymistry, and anatomy. Far be it from me to lessen now the importance and value of these divine sciences. They will all be useful, since luxury keeps pace with our medical knowledge ; for the violence and obstinacy, the variety and degrees of prevailing diseases have increased proportionally."

CHAPTER VI.

WHAT WAS ALCHEMY IN THE SEVENTEENTH CENTURY?

"I have observed that a reader seldom peruses a book with pleasure till he knows whether the writer of it be a black or fair man."—AD-DISON.

THE following paragraphs are taken from the title-page and preface of an old book by John French, supposed to have been among the last published works on alchemy in the English language. It was published in London, England, in 1650, and in repeated editions till 1657.

"The Art of Distillations : a full treatise of the choisest Sphagyrical Preparations in true *Alchymie ;* prepared by way of Distillation—with discriptions of the best Furnaces and vessels used by the ancient and modern Chemists. The anatomy of gold, and silver, and their choicest preper-ations, and their virtues ; in Six Books ; " " by John French Dr. in Physic." To which is added in the fourth edition, " Sublimation and Calcination, in Two Books." " Printed by E Cotes for F Williams, at the Bible in Little-britian, London, 1657." " Dedicated to my much Honoured Friend Tobias Gardbrand, Doctor of Physic and Princi-pal of Gloucester Hall, in Oxford, England." Dated in preface, Nov. 25, 1650.

" Nature and Art afford a variety of Sphagyrical Prep-arations, but they are yet partly undiscovered, and dis-persed in many books, and these of divers Languages, and partly reserved in private hands. When I consider what need there is of, and how acceptable, a general

THE ALCHEMIST.

Treatise of Distillations, and the Art of Alchymie might be, especially to our English Nation, I saw I could do no better service than to prepare a full Treatise which should contain only the choicest preperations, and only of the selectest Authors, both ancient and modern, and out of several languages, which I have attained, also some, by my own long manual experience, togeather with such as by exchange I have obtained out of the hands of private persons, which had been held as great secrets."

"I rejoice, at the break of day after a long tedious night, to see how this solitary Art of Alchymie, begins again to shine forth out of the clouds of reproach, which it hath a long time past undeservedly layed under. There are two things which have for a long time eclipsed it, *viz.:* the mist of unbelief, and Ignorance; and the specious Lunacy of conceit. Arise O Sun of truth, and dispel these interposed fogs, that the Queen of Arts may triumph.

"If man did but *believe* what this Art could effect, and what variety of wonder there is in it, they would be no longer bound up to Galen, or Aristotle, but would subscribe to be faithful to the principles of Hermes, and Paracelsus; as *they* still stand established, without Aristotle as their Prince, or Galen and Hippocrates as their Lords and Masters. They would no longer be dreaming forth *Sic dicet Galenus*, but, *Ipse dixit Hermes*. True, Galen and Hippocrates wrote excellently in many things. But that which I cannot allow is, their strict observance of the quadruplicity of humors, (which in the School of *Paracelsus*, and in the writings of *Van Helmont* hath been confuted), and their confining themselves to such simple crude medicines, which are more fit to be put into Sphagyrical vessels for *exaltation*, than into men's bodies to be fermented there."

"Certainly, if men were less ignorant, they would pre-ferre Cordial Essences, before Crude Jucies; Balsamic Elixers before flegmatic Waters, the Mercury of Philoso-phers before common Quicksilver. But many have so little faith in this Art, that they scarcely will hear any-thing about it, beyond distilling Waters, and Oyls, and extracting Salts; nay, many learned persons are so unbe-lieving that, as saith Sandivogius, if we should show them the Art, yet they would not by any means believe that there was any water in the Philosopher's sea. As for the posebility of the Elixer, you might as well try to per-suade them that they know nothing. Yet did not *Arte-sius* by the help of these medicines live 1000 years? Did not Flammel build fourteen Hospitals in Paris; besides as many in Bologne; besides Churches and Chappels with large revenues to them all? Did not Bacon do many miracles and Paracelsus very many miraculous cures? Thus much, by way of replying to the frivolous objections to the verity of this Art; and to those who will not even believe in it. If you should discover to them the very process of the Philosopher's Stone, they would only laugh at your simplicity, and will I warrent you never make use of it. Nay, if you should make projection before them, in their sight, they would think even then that there was a fallacy, so unbelieving are they, and so false learned. So I find them, and so I leave them, and shall forever. Then there is another sort of men by whom this art hath been scandalized, by carrying about and vending their *whites*, and their *reds*, their sophisticated Oyls, and Salts, and their ill-prepared *Aurum Vitæ*."

"Now we must consider that there are degrees, in this Art; for there is the accomplishing by *successions* of the Elixer itself, and there is the discovering of many essences, magisteries, and spirits, &c. Is not the *Ludus* of *Paracel-*

sus, that dissolveth the stone, and all salts and tartarous matter in the human body into a Liquor, worth finding out? Is not his *Tinca Scatura* a most noble Medicine, that extinguisheth all preternatural heat in the body *in a moment?* Is not his *Altahest* a famous dissolvent, that can in an instant, dissolve all things into their first principles? and withall is a *specificum* against all distempers of the Spleen, and Liver, worth finding out?

" A whole day would fail to recon up all the excellencis the sphagorical Art might bring to light, by *repeated* processes. In the searching out of which why may not the Elixer itself at last be attained? Is it not possible, by passing them through many Philosophical *repetitions*, to unfold at last the Riddle and Hieroglyphicks of the Philosophers? Is there no *fundamentum in re* for this? Is there no Sperm in Gold? Is it not possible to *exalt* it for multiplication? What is that which makes Gold incorruptible? What induced the old Philosophers to examine Gold for the best matter of their medicines? Was not Gold once living? Is there none now to be had, or did Sandivogius, the last of known Philosophers, use it all? There is enough if we do but find it out. If so, let no man be discouraged in the prosecution of it, especially if he keep the five Keys, which Nollins set down as a secret, and which all Philosophers with one consent enjoin the observation and use of."

Here follows the Theory and Practice, and the Prescriptions of the Alchemists. John French, the old author we quote, dedicated his works, as we have shown, to Tobias Grabrand thus: " Dear Sir, I once read of a nobleman's porter, who let in all that were rich apparelled, but excluded a poor Philosopher; but I should, if I had been in his place, have rather let in the Philosopher without the gay clothing, than the gay cloths without the

Philosopher.   As long as I have sence and reason, I shall
improve them to the honour of Art, especially that of
Alchymie.   In the perfection thereof there are riches,
honour, health, and length of days.   By it Artefius lived
1000 years.   Flammel we know built 28 Hospitals in
Paris and Bologne, with large revenues to them, besides
Churches and Chappels.   In the perfection of this Art,
I mean the accomplishing of the Elixer, is the mystery,
or Sulphor of Phylosophers, set at liberty; which grati-
fies the releasers thereof with three Kingdoms, *viz.* Vegitable, Animal, and the Mineral; and what cannot they do,
and how honourable are they that have the command
of these?   Did not Paracelsus many miraculous cures?
They may command without reason Lead into Gold; dying
plants into fruitfulness; the sick persons into health;
old age into youth; darkness into light; and what not."

"Court the Mother and you win the Daughter; prevail
with Nature and the fair Diana of the Philosophers is at
your service.   Now if you do not prevail for the fairest,
viz., the Sulphur or Mercury of the Philosophers, yet
Nature hath also other Daughters of wonderful beauty,
as the *mystic* Essences, and Magistrices of Philosophers;
which also are endowed with riches, honour, and health;
and some of these you may more easily try for."

"This art of Alchymie is the *Solary* Art, which is more
wonderful than all the other Arts and Sciences; and if you
did once make it shine forth, out of the clouds, whereby
it is eclipsed, it would eclipse and darken all others.   This
Queen of the Arts, is that true philosophy that most ac-
curately anatomises, and atomizeth Nature."

"Finally, when thou betakest thyself to the work of
'Practical Alchymie,' propound to thyself some one principle, and enter not upon it until thou art well versed in
thy theory; for it is better to work with the brain and

imagination, than with thy hands. Especially study nature well, and see if thy proposals are agreeable to the possibility thereof. Diligently read the sayings of *true* philosophers, read them over again and again ; and meditate on them ; and take heed thou doest not read the writings of others, instead of the true Books of the Philosophers. Compare their sayings with the possebilites of Nature. Compare the obscure parts or places with the clear, and where Philosophers say they have erred, do thou beware ; and consider well ' The general Axioms,' of Philosophers, and re-read so long till thou seest a sweet Harmony, and then Consent in their sayings. One thing further let me desire thee to take notice of, viz. whereas every process is set down plain, yet all of them, and others, must be proceeded in *Secundum Artem Alchymistæ.*"

Chaucer, in the " Canons Yeomans Tale," quotes a dialogue accredited to Plato, from a book called " *Senioris Zadith fil Hamuelis tabula chymica,*" whose subject is called forth by another alchymist work, popular in the middle ages, " *Secreta Secretorum,*" supposed to contain the sum of Aristotle's instructions to Alexander.

" ' For this science and this comming,' quoth he, ' is of the secret of secrets,' " concerning which he makes a dubious revelation under the name of Magnesia, and " spake so mistily," that it calls forth the new question from his disciple :

> " ' What is Magnesia, good sir, I pray ? '
> ' It is a water that is made, I say,
> ' Of th' elementës fourë,' quoth Plato.
> ' Tell me the rootë, good Sir,' quoth he tho,
> ' Of that water, if that it be your will.'
> ' Nay, nay,' quoth Plato, ' certain that I n'ill.
> ' The philosophers sworn were every one
> ' That they should not discover it to none.' "

And upon this the poet reads them a philosopher's moral
not "mistily," that God inspireth whom He will in this
knowledge, and he who goes contrary

> "' . . . . maketh God his adversary,
> ' As for to work any thing in contrary
> ' Of his will, certes never shall he thrive,
> 'Though that he multiply term of his live.
> ' And there a point ; for ended is my tale.
> 'God send ev'ry good man boot of his bale.'"

Old-time Popular Medicines.

" *To make the Magestry of Blood.* * —Take of the purest
blood as much as you please, put it into a Pellican, that
three parts of four may be empty, and digest it a month
in Horse-dung, (in which time it will swell and become as
much more as it was, when it was put in), then distill off
the fleghm in *Balneo*, and in the bottom will remain the
Magestry of Blood, which now must be distilled and co-
hobated *nine times*, in a Retort in ashes, and then it is
perfected.

"This Magestry, is of excellent virtue, which being
taken inwardly and applied outwardly, .easeth pains, and
cureth most diseases."

"*Elixer of Mummie.*—Take of mummie (viz. of Man's-
flesh hardened) cut small, four ounces. Spirits of Wine
terebinthenated, ten ounces, put them into a glazed vessel
(three parts of four being empty), which set in Horse-
dung to digest for the space of a month ; then take it out
and express it, let the expression be circulated a month,
then let it run through *Manica Hippocratis ;* then evaporate the Spirit, till that which remains in the bottom be

* " The Art of Distillation and Alchemy," by J. French, Doctor in Physic.
London, 1657 ; copied from pages 112 to 214.

Retzsch Del., 1824.

THE ART OF ALCHEMY.

like Oyle; which is the true Elixer of Mummie. This
is a wonderful *prevention* against all Infections."

"*The Essence of Mans-brains.*—Take the Brains of a
young man, that hath died a violent death, togeather with
its membranes, Arteries, Veins, Nerves, and all the pith of
the Back bone; bruise these in a stone mortar till they
become a kind of pap, then put as much of the Spirits of
Wine as will cover three or four fingers bredth: then put
it into a large glass, that three parts of four be empty,
being Hermetically closed, then digest it half a year in
Horse-dung; then take it out and distill it in *Balneo*, and
cohobate the waters till the greatest part of the Brains be
distilled off.

"A scruple, or a drop or two, of this Essence, taken in
some specifical water, once a day, is a most *infallable*
medicine against the Falling-sickness."

"*A famous Spirit of Cranium-humanum.*—Take human
craniums as many as you please, break them into small
pieces, which with water put into a glass-retort, well
coated, with a large Receiver well luted, then put a strong
fire to it by degrees, continuing it till you see no more
fumes come over; and you shall have a yellow Spirit, a
red Oyl, and a voltile Salt. Take this Salt and the yellow
Spirit, and digest them by circulation three months in
Balneo, and thou shalt have an excellent moist Spirit.
This is the same as that famous Spirit of Dr. Goddard's
in Holborn It helps Gout, Dropsie, infirm Stomach,
and indeed strengthens all weak parts, and openeth all
obstructions, and is a kind of *General Panacæa.*"

"*Another excellent Spirit of human sculls.* —Take of
human sculls without bruising them, only breaking them
into pieces, lay them piece by piece upon a net spread
over the wide mouth of any vessel, being almost full
of water, and cover this all with another vessel very

close but with vent, and then make the water boil and
keep it boiling three days and three nights; and in that
time the Bones will be as soft as cheese; then pound
them fine, and to every pound thereof add half a pound
of Hungarian Vitrol uncalsigned, and as much spirits of
Wine as will make it into a soft paste.   This paste digest
in a vessel Hemetrical sealed the space of a month in
Balnco, then distill in a Retort in Sand, till all be dry:
and you shall have a most excellent Spirit.   This in
very small doses is of wonderful use in the Epileptic Convulsions, in all Fevers putrid or pestilential, in passions
of the Heart ; and if taken in some Liquor it is an excellent Sudorifik."

"*Eye-water of Milk.*—Take of Womans-milk a pint, of
white Copperas a Drachm, distill them in ashes.   Note,
that as soon as thou perceiveth any sharpness to come
over, then cease.   Let inflamed eyes be washed with this
three or four times a day, for it cureth wonderfully."

Old-fashioned Tincture of Rhubarb, of the Alchemists.
—"Take Moncks Rhubarb, cut it in small pieces, one
ounce, pour thereon the Oyl of bitter Almonds, about
three fingers high, or four ounces ; set it in the heat of
the Sun, a Philosophical month, or 40 days, so that the
Oyl takes the Essence of the Rhubarb to itself : then
press it hard, and to that which is expressed, add spirits
of wine rectefied and leave it for some days in *Balnco
Mariæ*, so the spirit of wine doth attract unto it the
whole Essence; then put the remaining Rhubarb again
with more Oyl to digest until it be tinged ; and again
extract the Tincture with spirits of wine, which repeat
so often till the Rhubarb yields no more Tincture.   Then
distil off half the tinged spirits of wine, which return
again, and again, and then distil off half ; which work repeat four times, and at last distil off the whole together.

Thus is the Tincture brought over, *per Alembicum*, with the spirits of wine. Afterwards you must seperate it, and bring it to be like honey, in the form of a Balsam, in *Balneo Vaporoso*.

"Its dose is 10 drops, in some Conserve of Roses."

"*Aqua Splenetica*, famous for purefying the Spleen.— Take roots of the finger fern, and parsley, polyphody, lovage, birthwort, calamus, and acorns of the sea, of each an ounce; leaves of the scolopendria, wormwood, fumitary, dodder, agrimony, ceterach, viper grass, goats rue, harts-horn rasped, of each half an ounce; grains of paradise, red roses, lavender flowers of each a handful and a half; rich wine eighteen pounds; mix and let these digest 12 days, then distil off in *Balneo Mariæ*."

"This cutteth, discusseth, or chiefly purefyith the *Spleen;* and strengthneth the heart, and the head."

"*Restorative Liquor of Meat.*—Take the heart, lungs and liver of a good Calf, and the heart, lungs, and liver of a Fox, new killed, cut them all small, and add to them a quart of Shell Snails, after scoured in salt-water; then let them be put into a Copper vessel tinned withinside, and covered close that no vapour come forth; set this vessel over the steam of seething-water for 24 hours or thereabouts, and they will be for the most part of them turned into a Liquor of themselves; then take out this liquor and put it into a large Pellican, or Bolthead, putting to them two quarts of old Malligo-wine; Rosmary flowers, Marygold, and Marshmallows flowers, of each a handful; a half pound of raisins of the Sun, stoned, of mace or nutmeg a drachm; digest all together the space of a fortnight: then pour off that which is clear, from the feces, and sweten it with sugar, or syrup of Gilly-flowers, and let the Patient take five spoon-fulls three times a day. This recovereth the *decaying* strength

10

wonderfully: also useful for those who can neither eat nor digest food. It is very useful in Consumption, and repairs the radial moisture marvellously."

" *Oyl of Snakes and Adders.*—Take Snakes, and Adders equal parts, when they are fat, which is in June or July; cut off their heads, and tails, and take off their skins and unbowel them, and put them into a Glass gourd, and pour on so much of the pure spirits of Wine well rectefied, that it may cover them four or five fingers' breadth; stop the Glass well, and set it in *Balneo*, till all the substance be turned into an Oyl; which keep well stoped for use. This Oyl doth wonderful cures, in recovering hearing in those that be quite deaf, if a drop be nightly put into the Ears."

" *The Bears-balsam.*—Take of Bear's feet a pound, distil in a Retort and rectifie three times: then to this put the mother tincture of Saffron, Rosmary, Sage, and the spirit of Wine fortis, of each three ounces; mix well togeather and let remain in warm ashes for the space of a night; then strain and pour off the oyl and put four ounces of best pure beeswax while hot and mix quickly. This is an incomparable Balsam, to apply for Stiffness, the Gout, and Palsie."

" *The Quintessence of Snakes, Adders, and Vipers.*—Take of the biggest and fattest Snakes, Adders, and Vipers, which you can get best in June or July, cut off their heads, take off their skins and tails, and unbowel them, then cut them into small pieces, and put them into a Glass of a wide open mouth, and set them in a warm *Balneo*, that they may be well dryed, which will be done in three or four days; then take them out, and put them into a Bolt-head, and pour on them the best alcolizalid Wine, as much as will cover them seven fingers' breadth. Stop the Glass Hermetically, and digest them fifteen days in

Balnco, or so long till the Wine be completely coveted, which pour off; then pour on more of the aforesaid spirit of Wine, till all the quintessence be extracted. Then put all the tinged spirits together, and draw off the spirits in a gentle *Balnco*, till it be thick at the bottom ; on this pour spirits of Wine caryophilated, and stir them well to-geather, and digest them in a Circulatory ten days ; then abstract the spirit of Wine, when the quintessence re-maineth at the bottom perfect.

"This Quintessence, is of extraordinary strength and virtue for the purefying of the blood, the flesh, and the skin ; and consequently clenseth of all diseases therein. It cures also the Falling-sickness, by drop doses, and strengthens the Brain, Hearing, and Sight, and preserv-eth from Gray-hairs; reneweth the old to Youth, pre-serveth Women young, cureth the Gout, and Consump-tion; and is good aginst Stings, Bites, and Pestilential infections."

" *Pure Viper Wine.*—Take the best large and fat Vipers, four or six, according to their bigness, and put them into a gallon of the best Canary-sack, let them stand three months, then draw off as you take it. Some put the Vipers alive into the Wine, and there suffocate them, and afterwards treat them as above. This Viper Wine hath much the same virtues as the fore-going Quintes-sence; but it also provoketh to Love, and cures the Leprosy, and all worst corruptions of the Blood."

"*Aqua-Magnanimitatis;* the famous Water of Kun-rath.—Take of Pismires, and Ants, the biggest, (that have a soureth smell are the best) two handfuls, spirits of Wine a gallon : digest them in a Glass vessel, close stoped, the space of a month : in which time they will be dissolved into a Liquor, then distill them in *Balnco*, till all be dry. Then put the same quantity of Ants as before, digest, and distil them in the same said Liquor as before : do this three times, then aromatize the Spirit, with some aromatic. *Note*, that upon the spirit floats on Oyl, which must be seperated.

"This spirit is of excellent use to stir up the Animal Spirits : in so much that John Casmire, Palgrave of the Rhine, General against the Turks, did always take a little of it, when they went to fight, to increase courage and magnanimity, which it did to admiration. It doth also wonderfully wake up those any way slothful ; and the spirits that are dulled or dead with any cold distemper. This seperate Oyl, doth the same effect and indeed more powerfully ; it helpeth deafness by dropping it into the Ears ; this water helpeth also the Eyes that have any film growing on them, it being dropped into them."

"*Aqua-Magnanimitatis fortis.*—Take of Ants or Pismires a handful, of their Eggs two hundred, of Millipedes, i.e. Woodlice, one hundred, of honey Bees one hundred and fifty : digest these in two pints of spirits of Wine, being very well impregnated with the brightest soot ; digest them togeather for the space of a month, then pour off the mother tincture and keep it safe. This must be dilute with water or spirit, and is of the same vertue as the former."

"*Water of Amber*, made by Paracelsus, out of Cowdung.—Take of cow-dung, fresh, and distil it in *Balneo*, and the water thereof will have the smell of *Ambergriese*. This water in very small doses is excellent in all *inward* inflammations."

" *Water of Swallows*, wonderful against Fits, and the Falling-sickness.—Take of Swallows, cut into pieces the whole birds without seperating anything from them, six ounces; mix the meat togeather well, and add of Castorum cut small an ounce, then infuse them twelve hours in two pints of Canary wine, then put them into a Glass gourd, and distil them in Sand till all be dry, then cohobate the Liquor three times. This Wine taken in very small doses, in mornings, cureth them that have the Falling-sickness, or Fits, or the Staggers."

" *Water of Dung.*—Take of any kind of dung as much as you please, whilest it is fresh, put into a Common Still with white Wine enough to moisten it, and with a slow or soft fire distil it off : it will be better if the Still be over a water-vapour, and if thou wilt have it stronger, cohobate the water or wine over the fires many times repeated, for we see there is great vertue in dung, and many sorts thereof are very *medicinable.* If we take Doves-dung, the water is very excellent for obstruction of the Kidneys, Bladder, or Liver, helpeth the Jaundice presently. If Horse-dung is sphagericaly treated in like manner, it is good against the Bastard-plurisie, Stitches, Wind, Obstructions, Dropsie, Scurvy, &c."

" *Water of the Sperm of Frogs.*—Take of the Frog-spawn gathered in March, about the new of the moon, four pounds, Cow-dung fresh six pounds, mix them well togeather, and let them stand the space of one day, then distil them close in ashes. This water allays all hot pains, both inward, and outward, in fevers, and especially in Gout."

" *Water of Meat.*—Take what Flesh you please, the bloodiest parts, unwashed, cut very fine, then bruised, (or if it be a feathered fowl, take it after being chased up and down until it be weried, and then suddenly strangled, the feathers then plucked off without putting it into water, and the bowels taken out and it all clean, cut the flesh, bones, gizzard, liver and heart) and pour upon it as much cold water as sufficient, with spices and herbs, then set it over a gentle fire in an earthen glazed vessel, for the space of 24 hours; then put on the head and lute it close, and there will distill off a delicious restorative Water of Meat; good for the feeble."

———

" *To Fortify a Load-stone.*—Take a Load-stone and heat it very hot in coals, but so that it be not fired, then quickly and perfectly quench it in the Oyl of *Crocus Martis*, made of the best Steel, that it may imbibe as much as it can. Thou shalt by this means make the natural Load-stone so very strong and powerful, that thou mayest pull out nails out of a piece of wood, with it; and besides do wonderful things with it. Now the reason of this is, as saith Paracelsus, ' because the *spirit* of Iron is the life of Load-stone, and this may be extracted from, or increased in the Load-stone.' "

" *Luminous Water.*— Take the tails of many Glow worms, put them in a Glass still, and distil them in *Balneo;* then pour the said water upon more fresh tails of Glow worms; do this fifteen times, then thou shalt have a most Luminous Water, by which thou maist see to read in the darkest night."

" *The World in a Glass.*—Take of the purest Salt Nitre as much as you please, of Tin half so much; mix these well togeather, then calsigne them Hermetically; then put them into a Retort, to which annex a Glass-Receiver, and lute them well togeather, let there be leaves of Gold put into the bottom thereof, then put fire to the Retort, until vapors arise that will cleave to the Gold : augment the fire till no more fumes ascend ; then take away the Receiver and close it Hermetically, and make a Lamp-fire under it, and now you will see represented in it, the Sun, Moon, Starrs, Fountains, Flowers, Trees, Fruits, and indeed all things, which is a glorious sight to behold."

" *The Golden Mountain.*—Take of snake Adders' Eggs, half a pound, put them into a Glass Retort, distil them by degrees ; when all is dry, you shall see the feces at the bottom turgid, and puffed up, and seem to be, as it were, Golden Mountains, being very glorious to behold."

" *Dr. Burges* his *Plague Water.*—Take three pints of Muscadine, and boyle in it Sage and Rue, of each a hand-ful, till a pint be wasted ; then strain it, and set it over the fire again, put thereto a drachm of long-pepers, of Ginger, and Nutmeg, each half an ounce, being all bruised togeather ; then boyl them a little, and put thereto half an ounce of Andromachus-treacle, and three drachms of Mithridate, and a half pint of the best Angelica-water. This must be kept as your life, and above all earthly treas-ure ; and must be taken to the quantity of a spoonful morning and evening, if you be already infected, and sweat thereupon. If you be not infected, a spoonfull is suffi-cient, half in the morning, and half at night. All the Plague-time, under God, trust to this ; for there was never a man, woman, or child, that ever failed of their expecta-tion in taking it :"—more faith than medicine in such emergency !

" *A Water that purgeth, without any pain or griping.*—Take of Scammony one ounce, Hermodactils, two ounces ; the seeds of Broom, lesser-Spurge, Dwarf-elder of each

half an ounce : the juice of wild Asses cucumber, of black
Hellebore, the fresh flowers of Elder, of each an ounce
and a half : Polypodium, six ounces ; of Senna three
ounces ; Red-sugar eight ounces ; distilled water six pints.
Let all be bruised and then infused in the water 24 hours,
then distilled in *Balnco*. This water may be given from
one drachm to two ounces ; it purgeth all manner of humors, opens all obstructions, and is pleasant to be taken :
they whose stomach loathe all other things, may take
this acceptably. Notice, there may be hanged a little
bag of Spices in this distilled Water, and it may be swet-
ened as taken."

" *Liquor against the Tooth-ache.*—Take of Oyl of Cloves
half an ounce, dissolve in it half a drachm of Camphor :
add to them half an ounce of spirits of Turpentine, and
half a drachm of Opium. A drop or two of this Liquor
put into a hollow tooth with some lint, easeth the toothache presently.

" *Oyl of Amber*, against all dreadful diseases.—Take
of light Amber one part, of the powder of flints calsigned,
and of tiles powdered two parts ; mingle them, and put
them into a Retort, and distil them in sand. The Oyl
which is white and clear, which is first distilled off, keep
it by itself ; then continue the process as long as any Oyl
distils off. The salt of amber, which, you see, adheres to
the neck of the Retort within side, being gathered, let it
be purified by solution, filtration, and coagulation, according to art, and be kept close for use. After the
process is ended, and all be cold, let there be a rectifica-
tion and seperation of the clear Oyl, from the fœtid Oyl,
after this manner. Put the distilled Liquors into a glass
body, and distil it in *Balnco Mariæ*, with a fire strong
enough, and first the Phlegm or spiritual part will distil
over, and also the Golden Oyl, and swim on the Liquor,

which is then to be seperated by a glass tube, and kept for use. But the strong blackish Oyl that remains in the stil, keep by itself—a Balsam of amber.

" This yellow Oyl of Amber was formerly esteemed Sacred. Its excellent virtue, is effectual in the epilepsie, Apoplexie, Meloncholie, Cramps, Vertigo, Stone, Pestilence, Cold defluxions in the head, Pain or palpitation of the Heart, and for such as are troubled in their minde ; for the Jaundice, difficulty of Breathing, difficulty of the Menstruums, the white flux, difficulty of urine, hard Travel, Strangulation of the Womb: for Fevers, and for Worms. The dose is from one to two grains. The Balsam of Amber is used externally or internally for the suffocation of the Matrix. It is also used as emplasters for closing wounds."

" *Usque-bath*, or the *Irish Aqua vitæ* is made thus. Take a gallon of spirits and put therein a quart of Canary Sack, and all in a Glass-vessel ; add two pounds of Raisins of the Sun, stoned, but not washed ; two ounces of Dates stoned, four nutmegs, and an ounce of extract of liquorice, all bruised togeather coarsly; stop the vessel very close and let them infuse in a cold place for seven days; then let the Liquor run through a bag called *Manica Hippocratis*, made of cotton. This Liquor is used for the Stomach after surfeits."

" *A Compound Oyl, against the Suffocation of the Matrix, and the Megrim.*—Take of Rue powdered, one pound; Castorium two ounces, Olibanum and Myrrh of each four ounces ; Linceed Oyl one pound and a half. Let them be digested togeather for fourteen days, in *Ventre Equino*, or such like heat while close stoped ; afterwards distil it by a Retort in a close Reverberatory. With this Oyl anoint the Navil morning and evening ; or the forehead and temples, or both the Navel and Temples."

" *To Discover* what kind of metal there is in any sample, substance, or Ore, though you have but a few grains thereof.—Test: Take 2 to 4 grains, or a little more if you have it, of any Ore, or substance, put it to half an ounce of fine broken Venice-glass, and melt them together in a Crucible over a fire of charcole, it being covered, and according to the tincture or colour that the glass receiveth, so may you judge what kind of metal there is: for if it be a Copper Ore, then the glass will be tinged with a Sea-green colour.

"If Copper and Iron, a Grass-green.

"If Zinc, a golden-purple.

"If Iron, a dark yellow.

"If Tin, a pale yellow.

"If Silver, a whitish yellow.

"If Gold, a fine golden *skie* colour.

"If Gold and Silver togeather, a Smaragdine colour.

"If Gold, Silver, Copper, Zinc, and Iron togeather, an Amethist colour."

" *Hepar Sulphor*, the same as Crocus Metallorum; or the Liver of Antimony.—Take crude Antimony, and of Salt Peter each a like quantity; beat them small, mix them and put them into a strong Iron Mortar, inclined sidewise; then kindle the powders with a strong *quick* Char-cole fire under, or a red-hot Iron-rod put into them, and so will the Antimony be fixed, and seperated from its Arsenical Sulphur; then seperate the Salt Peter from it, and further edulcorate the real *Hepar Sulphor*."

" *The Spagirical Anatomy of Gold.**—I shall first show whence Gold had its origin, and what the matter thereof is. As Nature is in the will of God, and God created her, so Nature made for herself a seed ; i.e. her will in the elements. Now she is indeed one, yet she brings forth

* Dr. French, "Chemistry and Alchemy," book vi., page 189.

divers things; but she operates nothing without a Sperm.
Whatsoever the Sperm willeth, Nature operates, for she
is, as it were, the instrument of any Artificers. The
Sperm therefore of every thing is better, and more prof-
itable than Nature herself; for thou shalt from Nature
without Sperm do as much as a Goldsmith without a
fire; or a Husbandman without grain or seed. Now the
Sperm of anything is the Elixer, the Balsam of Sulphor,
and the same as Humidus Radicale is in metals.

"Four Elements generate a Sperm, by the will of God,
and imagination of Nature . . . . so the four Elements
by their indefinent motion, each according to its quality,
casts forth a Sperm into the centre of the Earth, where
it is digested, and then by motion is sent abroad. Now
the centre of the Earth is a certain empty place, where
nothing can rest; the four Elements send forth their
qualities, to the circumference of the centre . . . . so it
occurs in the centre of the Earth, that the magnetic
power of a certain part, or place, attracts something convenient to itself, for the beginning and bringing forth of
some body, and the rest is cast forth as stones and other
excrements. For every thing hath its original from this
Fountain.

"For example, pour upon an even Table some water,
in the middle thereof, and round about the water place
divers things and divers colours thereof, and Salts, &c.,
but every thing by itself near the edge of the water.
You shall see as the water touches the red color, it will
be made red by it; if the salt, it will attract and be saltish, and so the rest. Now the water doth not change
the things, but the diversity of things, each by its own
nature so changeth the water. In like manner the seed
or sperm being cast forth by the four elements from the
centre of the Earth unto the superficies thereof, passeth

through various places, and according to the nature of
the given place, is any thing produced : if it come to a
pure place of the earth, then water, a pure thing is made.
So you see that seed and the Sperm of all things is but
one, and yet it generates divers things, as shown by the
foregoing example.   The Sperm, while it is in the centre,
is indifferent to all forms, but when it comes into any
determinate place, it changeth no more its form.   The
sperm whilst it is in the centre can as easily produce a tree,
as a metal, or an herb, as a stone, and one more precious
than another, according to the richness or purity of the
place.   Now this sperm is produced of Elements thus:

"The four Elements are never quiet but by reason of
their similarity, or by their contrariety, and mutually acting one upon the other; and every one of itself sends
forth its own subtlety, and they agree then in the centre.
Now in the centre is the Archæas, the servant of Nature,
which mixes those sperms togeather, sends them abroad,
and by distillation sublimes them, by the heat of continual motion, unto the superfices of the Earth.   For the
Earth is porous, and this vapour or wind, or aura, as the
Philosophers call it, is by distilling through the pores of
the earth resolved into water, *of which all things are produced.*   Let all the sons of Art know, therefore, that the
sperm of metals is not different from the sperm of all
things else ; viz., it being a humid ether vapour, or aura.
Therefore in vain do Artists endeavour the reduction of
metals into their first matter form, which is attenuated
and only imperceptible.   When Philosophers speak of
first matter, saith Bernard Trevisan, they did not mean
this vapour, but the second matter, which is unctuous
water; which to us is the first, because we never find the
former."

"The specification of this vapour into distinct metals,

is thus: This vapour passeth in its distillation through
the earth, through places either cold, or hot. If through
a hot and pure place where the fatness of Sulphor sticks
to the sides thereof, then that vapour, which is called the
Mercury of Philosophers, mixeth and joineth itself to the
fatness, which afterwards it sublimes with itself; and
then it becomes, leaving the name of a vapour, an Unc-
tuosity, which the antecedent vapour did purge, where
the earth is subtle, pure, and humid, and fills the pores
thereof and is joined to it ; *so it becomes Gold ;* but where
it is hot and yet something impure, Silver. But if the
fatness comes to impure places, which are cold, it is made
Lead ; but if that place be pure and mixed with sulphur
it becomes Copper; for by how much the more pure,
and warm, the place is, so much the more excellent doth
it make the metals, even to pure Gold." . . . . "Perfect
Health, and pure Gold are alike. Geber also asserts the
same thing, when he saith, that all sorts of sick, or imperfect bodies, of whatever nature, have superfluous hu-
midities, prone to generate further combustible crudity
and corruption, and some have an impure, feculent natural grossness."

" Now if any skilful Philosopher could wittily seperate
this adventitious impurity from gold, while it is yet living
(for from common gold it never can be by reason of the
Spirits that are bound up and as good as dead in it) he
would set subtle Sulphor at liberty, and for this service
he should be gratified with three Kingdoms, viz. Vegetable, Animal, and Mineral ; I mean he could remove the
great obstacle which hinders gold from being digested
into the Elixer. For as saith Sandivagius, the Elixer, or
Tincture of Philosophers is nothing else but Gold, *digested*,
or attenuated unto the highest *potency* or degree. This

would be the Sperm for curing all diseases, and for making Gold *ad infinitum.*"

" If Gold consists of Mercury and Sulphor, as Paracelsus affirms, and if all Murcury can be reduced into a transparent water, which is one of the greatest secrets I know, or care to know, why may not that water, when well attenuated, in some sense, be called a kind of Living Gold, which will make perhaps a medicine, or *Menstruum,* unfit for the vulgar to know."

After several pages * occupied with numerous formulæ or prescriptions containing gold, as " Dr. Anthony's Aurum Potable," so famous in Italy, Germany, and Spain, also various oils and tinctures of gold, and other preparations of gold with mercury, is the following :

" *The Virtues of the aforesaid preparations of Gold.* With these medicines of Gold the Ancients did not only preserve the health and strength of their minds and bodies, but also prolonged their lives, to a very old age ; and not that only, but they cured thoroughly Madness, Melancholy, Apoplexie, Epilepsie, Pluresie, Leprosie, Lues Venera, the Wolf, Cancer, Consumption, Noli me tangere, Asthma and inward Imposthumes, and such like diseases, which most physicians account incurable. For there is such a *potency* of fire residing in prepared Gold, which doth overcome ; and not only quell or consume deadly diseases, or humors, but also renews the very marrow of the bones ; and raiseth the whole human body, though it had been half dead.

" Let old men take it twice a month, or weekly, for by this means will their old age be fresh ; till the appointed time of death.

" Let it be given often to those women who have past the years. . . . .

* French's " Alchemy," book vi., pp. 198–210.

" Let it be given to women in travil. . . .

" Let young men take any of these preperations of Gold, monthly, and they may expect long life.

" Let young women and maids take it once a month, and they will appear fresh and beautiful."

" I have set down several tinctures of Gold, and now I wish to give some more of the true Philosophers Gold and Silver; for indeed the Art of preparing these is the true Alchymy, in comparison of which all the new Chymical discoveries are but abortive attempts; else found out by accident. What unworthiness God saw in Gold, more than in other things, that he should deny the *seed* of multiplication to it, and give it to others, is hard to discover: more than to discover the Elixer itself. In trying, says *Sandivagius*, ' they try most difficult operations, and very subtle experiments and discoveries, which the Philosophers themselves never dreamed of.' 'Nay,' saith the afore named author, ' if *Hermes* himself were now living, togeather with subtle-witted *Geber*, and most profound *Raimund Lullie*, they would be accounted by our modern Cheymists not for Philosophers, but rather for learners.' They were ignorant of these so many Distillations ; so many repeated Circulations; so many Calcinations ; and so many minute and other difficult Operations of Artists, now a day used: which indeed men of this age did find out, and added to their Books ; Yet there is one thing wanting to us, which they did, viz., to know how to make the Philosophers stone, or *final* Tincture ; the process of which, or one according to some Philosophers, are these—

" *The Process of the Elixer, according to Paracelsus.*—Take the mineral *Electrum*, being imature, and made very subtle, put it into its own sphere, that the impurities and superflueties may be washed away then purge it, as much

as possibly you can, with *Stibium;* after the Alchymis-
tical way, lest by its impurity thou suffer prejudice.
Then resolve it in the stomach of an *Estridge,* which is
brought forth in the earth, and through the sharpness of
the *Eagle,* where it is comfortated in its vertue.

"Now when the *Electrum* is consumed, and hath, after
its solutions, received the color of a Mary-gold, do not
forget to reduce it into a spiritual transparent essence,
which is like to true Amber; then add half so much as
the *Electrum* did weigh, before its preperation, of the
extended Eagle, and *oftentimes* abstract it from it in the
stomach of the Estridge, and by this means the *Electrum*
will be made more *dynamic* or spiritual. Now when the
stomach of the Estridge is wearied with labour, it will
be necessary to refresh it, and always again to abstract
it. Lastly, when it hath again lost its sharpness, add the
Tartarized Quintessence; yet so, that it be spoyled of
its redness the higth of four fingers, and that must pass
over with it; this do so often till it be of itself white,
and when it is enough, and thou seest that sign, sublime
it. So will the *Electrum* be converted into the whiteness
of an *exalted* Eagle, and, with a little more labour, be
transmuted into a deep redness, and then it is fit for
Medicine."

"The Process of ' *The Elixir,*' according to *Divi Seschi,*
and *Pontanus.*—Take of our earth through *eleven degrees,*
eleven grains; of our Gold, and not of the vulgar, one
grain; of our *Lune,* not of the vulgar, two grains; but
be thou admonished that thou take not of the Gold and
Silver of the vulgar, for they are dead, but take ours
which are living; then put them into *our fire,* and there
will thence be made a dry Liquor. First the earth will
be resolved into water, which is called the Mercury of
Philosophers, and in that water it will resolve the bodies

of the Sun, and Moon, and consume them; that there remain but the *tenth* part, with *one* part, and this will be the *Humidum Radicale Metallicum.* Then take the water of the Salt Nitre of our earth, in which there is a living stream, if thou diggest the pit knee deep; take therefrom the water of it, but take it clear, and set over it that *Humidum Radicale*, and put it over the fire of putrefaction and generation, but not such as was in the first operation.

"Govern all things in this process with a great deal of discretion, until there appear colours like to the tail of a Peacock. Govern it by repeated digestions of it, and be not weary till these colours cease; and there appear throughout the whole, a green colour; and so of the rest: and when thou shalt see in the bottom, ashes of a fiery colour, and the water almost red, open the vessel, dip in a feather, and smeer over some iron with it, see if it tinge, having in readiness that water which is the *Menstruum* of the world, (out of the sphear of the Moon, so often rectefied, until it can calcine Gold,) put in so much of that water, as was the measure of cold air which went in: boyle it again with the former fire until it tinge again.

"Take the matter now and grind it by hand with a Physical *potential* contrituration, as diligently as may be done: then set it aside upon the fire, and let the portion of the fire be known; viz, that it only stir up the matter; and in a short time, that fire, without any other laying on of hands, will accomplish the whole work, because it will putrefie, corrupt, generate and perfect, and bring to appear the three principal colours, Black, White, and Red.

"So, by the means of our fire, the Medicine will be multiplied, not only in very quantity but also in *vertue.*

11

Withal, they might therefore try to search out this fire, which is invisible, equal, continual, vapours not away if so very infinite, except it be too much stirred up; it partakes of hidden Sulphur; is taken from elsewhere, not from the matter, but putteth down all things; dissolveth and neutralizeth all corruptions, congealeth, and calcines; and is artificial to find out, or to be understood, and that by a compendious way; is not transmuted with the matter, because it is not of the matter; and by it thou shalt attain thy wish, because itself doth the whole work; and is the very key of the Philosophers, which they know how to use, but never reveal."

" *The Smaragdine Table of Hermes,* whence all Alchymie did arise. True, without any falsity; certain, and most true. That which is inferior, is as that which is superior; and that which is superior, is as that which is inferior, for the accomplishing of the miricles by *one*, least thing. And as all things were first from *one*, and by the mediation of one; so all things have proceeded from *one*, by adaptation. The Father thereof is the Sun; the Mother thereof is the Moon; the Wind carried it in its belly; the Nurse thereof, is the Earth. This is the perfection of Art. Thou shalt always seperate the earth from the fire, the subtle from the thick, sweetly, with great judgment. So that all obscurity flees from thee, and thou doest wonderful adaptations. Hence, having three parts of all Philosophy, I am called *Hermes Trismegistus.*"

" *Homunculus,* the famous *Arcanum,* or Restorative Medicament of *Paracelsus.**

" First, we must understand, that there are three acceptations of the word Homunculus, in Paracelsus, which are these:

" 1. *Homunculus,* is a superstitious Image, made in the

* French's " Alchemy," book vi., p. 139.

place or name of any one, that it may contain an astral and invisible man; wherefore it was made for a superstitious use.

"2. *Homunculus* is taken for an artificial man, made of *Sperma*—digested in the shape of a man, and then nourished and increased with the essence of mans blood: and this is not repugnant to the possebility of Nature, and Art. But is one of the greatest wonders of God which he ever did suffer mortal man to know. I shall not here set down the full process, because I think it unfit to be done, at least to be divulged; besides, neither this, nor the former, is for my present purpose.

"3. *Homunculus*, is taken for an excellent Arcanum, or *Medicament;* extracted by the spagyrical Art, from the chiefest staff of the natural life in man, and according to this conception, I shall speak of it. But before I show you the mysterious process, I shall give you an account why this Medicament is called Homunculus; and it is this—

"No wise man will deny, that the staff of life is the nourishment thereof; and that the chiefest nutriment is Bread and Wine: being ordained of God, and Nature, above all other things for the sustenance thereof. Besides, *Paracelsus* preferred this nutriment for the generation of the Blood and Spirits, and the forming thence the sperm of his wonderful *Homunculus.* Now by a suitable allusion, the nutriment is taken for the life of man, and especially because it is transmuted into life; and again, the life is taken for the man; for unless a man be alive, he is not a man, but the *carkas* only, of a man, and the baser part thereof; which cannot perfectly be taken for the whole man, as the nobler part may. In as much therefore as the nutriment, or aliment of life, may be called the life of man, and the life of man be called man:

this nutriment extracted out of Bread, and Wine, and being by many repeated digestions *exalted*, into the highest purity and *potency* of a nutretive substance, and consequently becoming the life of man, being so *Potentially*, it may metaphorically be called *Homunculus.*

"The process, which in part shall be set down allegorically, is thus: Take the best Wheat, and the best Wine, of each like quantity, put them into a Glass, which thou must Hermetically close up; then let them putrefy in Horse-dung three days, or until the Wheat begins to germinate, or to sprout forth, which then must be taken forth, and bruised in a mortar, and be pressed through a linnen cloth, and there will come forth a white Juice like milk; you must cast away the feces, and let this juice be put into a Glass, which must be not above half full; stop it close, and set it in Horse-dung, as before, for the space of fifty days. If the heat be temperate, and not exceeding the natural heat of a man, the matter will be turned into a *spagyrical* blood and flesh, like an Embryo. This is the principal, out of which is generated a twofold sperm, viz, the father and mother, generating the *Homunculus*, without which there can be no generation." . . . .

"From the blood and flesh of this Embryo, let the water be seperated in *Balneo*, and the air in ashes, and both be kept by themselves. Then to the feces of the latter distillation, let the water of the former distillation be added, both which must putrefy in *Balneo*, the space of ten days; after this, distill the water repeatedly (which is then the *vehiculum* of the fire) togeather with the spirit, in ashes; then again distill off this water in *Balneo*, and in the bottom remains the fire which must be distilled in ashes. Keep both these apart; and thus you shall have the four Elements, seperated from the Chaos of the Embryo.

"Again, the feculent earth is to be reverbrated in a close vessel for the space of four days. In the interim, distil off the fourth part of the first distillation in *Balneo*, and cast it aside ; the other three parts distil in ashes, and pour it again into the reverbrated earth and distil it in a strong fire ; cohobate it four times, and so you shall have a very clear water, splendid, odoriferous, and must be kept apart. After this, pour again the spirit upon the first water and putrefy them togeather in *Balnco* the space of three days, then put them into a Retort, and distil them in Sand, and there will come over a water tasting of spirit : let this water be distilled in *Balnco*, and what distills off keep by itself, as also what remaining in the bottom, which is the fire, keep by itself. This last distilled water pour again upon its earth, and let them be macerated togeather in *Balnco* for the space of three days, and then let all the water be distilled in Sand, and let what will arise be seperated in *Balnco*, and the residue remaining in the bottom be reserved with the former residence. Let the water be again poured upon its earth, and then abstracted, and seperated as before, until nothing remains in the bottom which is not seperated in *Balnco*.

"This being done, let the water which was last seperated be mixed with the residue of its fire and be macerated in *Balnco* three days, and all be again distilled in *Balnco*, that can ascend with that heat, and let what remains be distilled in ashes from the fire and what shall be elevated shall be aerial ; and what remains in the bottom is fiery. These two last Liquors are ascribed to the two first principles ; the former to Mercury, the latter to Sulphor, and are accounted by *Paracelsus*, not as elements, but their *vital parts ;* being, as it were, the natural Spirits and Soul, which are in them by nature.

" Now, both are to be rectefied togeather and reflected into their centre with a circular motion, that this *mercurius* may be prepared with its water of *attenuation*, being kept clear and odoriferous in the upper place, and put the Sulphur by itself. Now it remains, that we look into the third principle : so let the reverbrated earth, being ground upon a marble, imbibe its own water, which did above remain after the last seperation of the Liquors made in *Balneo*, so that this be the fourth part of the weight of its earth, and be congealed by the heat of ashes into its earth, and let this be done repeatedly and so oft, the proportion being observed, until the earth has drunk up all its water.

"And lastly, let this earth be sublimed into a white powder, as white as snow ; the feces being cast away. This earth being sublimed, and freed from its obscurity, is the true Chaos of the first Elements ; for it contains those things *occult*, seeing it is the Salt of Nature, in which they lye hid ; being, as it were, reflexed in their centre. This is the Third principle of *Paracelsus*. It is the Salt, which is the Matrix, in which the two former sperms, viz, of Mercury, and Sulphur, are to be put, and to be closed up togeather in a glazen womb, sealed with *Hermes* seals, for the true generation of the *Homunculus* produced from the spagyrical Embryo : and this is the true *Homunculus*, or great Arcanum ; otherwise called the nutritive *Medicament* of *Paracelsus*.

"This *Homunculus* is of such vertue, that presently after a very little of it is taken into the body it is turned into Blood and Spirits ! If then diseases prove mortal, because they destroy the Spirits, what mortal disease can withstand such a medicine, that doth so soon repair, and so strongly fortifie the Spirits, as this *Homunculus ;* it being as Oyl to the vital flame, into which it is *immediately turned*, thereby renewing the same. By this medicament diseases are surely overcome, and *expelled;* so also is youth renewed, and gray hairs prevented."

" *The Spagyrical Anatomie of Water,*" (in 1650.) Water seems to be a body so very Homogeal, as if neither Nature or Art could discover any Heterogeneity in the parts thereof. Thus indeed it seems to the eyes of the Vulgar, but to a Philosopher far otherwise: as I shall make credible, by presenting to your consideration a twofold process, for discovering the dissemilar parts thereof; whereof the one is natural, only, and the other artificial. But before I speak of either, it must be premised, that in the elements of water there is great plenty of the Spirits of the World, which is more predominant in it, than in any other element, for the use and benefit of universal Nature ; and that this Spirit hath three distinct substances, viz, Salt, Sulphur, and Mercury. Now, by Salt, we must understand a substance very dry, vital, and radical ; having in it *the beginning* of corporification, as I may so call it. By Sulphur, a substance full of light, and vital heat, or vivifying fire, containing in itself the beginning of motion. By Mercury, a substance abounding with radical moisture, with which the Sulphur of life, or the vital fire is cherished, and preserved.

" Now these substances which are in the Spirit of the World, make all Fountains, and Waters, but with some differences, according to the predominancy of either. This several predominancy thereof, is the ground of the variety of productions, because all things are produced out of water; for Water is both the Sperm, and the *Menstruum* of the World ; the former, because it includes the seed of everything; the latter, because the Sperm of Nature is putrefied in it; that the seed included in it·

should be actuated, and take upon it the divers forms of things : and because by it the seed itself, and all things produced of seed grow, and are increased. Now this being premised, I shall show the natural process, which shall be made plain, in three productions, viz, the Spawn of Frogs, of Stones, and of Vegitables."

" The spawn of Frogs is produced in this manner ; viz. The Sulphur which is in the Water, being by the heat of the Sun resolved, and dissolved, is greedily, and with delight conceived by the element of water. . . . The water wants ficcity, which the Sulphur hath, and therefore exceedingly desiring it, doth greedily attract it to itself. Sulphor also wants humidity, and therefore attracts the humidity of the water. Moreover the humidity of the water, hath the humidity of the Salt laid up *occultly* in it ; also the Sulphur cherisheth the ficcity of the fire, and desires nothing more than the humidity of the Salt, that is in the water. Sulphor also contains the ficcity of the salt, whence it is that Salt requires a ficcity from the Sulphor. And thus do these attractive vertues mutually act upon each others subject. Now by this means there is a conception made in the water which now begins to be turgid, puffed up, and troubled, as also to be grosser, and more slimie, until out of the spermatic vesseles, the Sperms be cast upwards, in which Sperms, after a while, appear black specks, which are the seed of the Frogs, and by the heat of the Sun, are in a short time turned into the embryo of the same ; by which it appears there are dissimilary *parts* in water."

Of course not a word of the foregoing is true, but this was supposed to be a wonderful analysis of water by the alchemist in 1650. But further :

" 2d. Stones are produced out of water that have a mucilaginous Mercury, which the Salt, with which it also abounds, fixith into stones. This you may see clearly by putting stones into water; for they will after a time contract a mucilaginous slime matter, which being taken out of the water and set in the Sun shine, becomes to be a stony nature. And whence comes those stones, gravel, and sand, which we see in springs? They are not washed down out of the Mountains and Hills as some say, from whence the water springs, neither were they in the earth before the springs break forth, as some imagine, or if you dig round about the springs even beyond the heads of them, you shall find no stones at all in the earth, only in the veins thereof through which the water runs. Now the reason of the smallness of these stones is the continual motion of the water, which hinders them from being united into a continued bigness. I shall make a further confirmation of this in an artificial process of manifesting the heterogeneity of water; here only adding the assertion of *Van Helmont*, 'that with his *Altahest* all stones, and indeed all things may be turned into Water.' If so, then you know what the Maxim is, viz, All things may be resolved into that from whence they had their beginning.

" 3d. Vegitables are produced .out of water, as you may clearly see, by the water sending forth plants that have no roots fixed in the bottom; of which sort is the herb called Duck-weed, which putteth forth a little string, into the water, which is as it were the root thereof. For the confirmation of this, I will state, that a Gentleman in this city, London, at this time, of no small worth, saith he had fair water standing in a Glass, divers years, and at last a plant grew out of it. Also, if you put some plants, as Water-mint, &c. into a Glass of fair water only, it will germinate, so that the roots shall fill the Glass. Hereunto may be added the experiment of *Van Helmont*, concerning the growth of a tree; for saith he—

"I took two hundred pounds of earth by weight, dryed in an oven, and put into a vessel, into which I set a Willow-tree which weighed five pounds; which by the addition of water to the earth did in five years time, grow to such a bigness, as that it weighed one hundred and sixty nine pounds; at which time I also dried and weighed the earth, and within two ounces it retained its former weight. Besides, the Ancients have observed that in some places herbs have grown out of the snow; and do not we see that all Vegitables are nourished, and increased, with insipid water; for what else is their juice?

"If you cut a Vine in the month of March, it will drop gallons of water, which insipid water if it had remained in the trunk of the Vine, would in a little time have been digested, into Leaves, Stalks, and Grapes, which Grapes also by further maturation would have yielded Wine; out of which you might have extracted a burning spirit, which spirit would flash. Now, I say, although this insipid water, be by the specifical Sulphur and Salt of the Vine, fixed into the Stalks, Leaves, and Grapes of the Vine, yet, these give it not a coporificative matter, for that it had before; or an unseen aptitude and potency to become what afterwards it proved to be: for indeed Stalks, Leaves, and Grapes, were *potentially* in it before; all of which now becomes seen to be actual, by virtue of the Sun, and of the aforesaid Sulphur and Salt, whereof the two latter were actually in the *very small* seed; and therefore, as I said, could not add any bulk to them.

"Moreover, do not we see, that when things are burnt and purefied, they ascend up into the air by way of vapour and fumes, and then descend by way of insipid dew, or rain? Now what do all these signifie, but that from water, are all things produced, and in it are dissimilary *parts?*

" The Artificial process is this. Take of what water you please, as much as you please; let it settle until quite clear. Then digest it the space of a month; after which time evaporate the fourth part, by a very gentle heat, and cast it away, it being but the phlegm; then distil off the remainder of the water, till the feces only are left; which will be a slimy saltish substance. This middle substance distil again, as before, casting away every time in succession the fourth part, and keeping the remainder by itself for further use; and this repeat seven times. Note, that after the fourth or fifth distilation, the water will distil over like Milk, colouring the head of the Still, so that it can hardly be washed or scoured off.

"This water, after the seventh distillation, will leave no feces behinde; and if you digest it for three months, it will be coagulated into Stones and Crystals, which some extol very much for a Medicine, to cure inward and out-ward putrefactions; out of which also may be made a dis-solving Spirit. Note, that as this resultant water stands in digestion, you may see divers live colours. Now as for the feces, which I spoke of, which indeed all waters, even the swetest leave at the bottom, being as I said a saltish slime, and in taste as it were a medium between salt and nitre : take them and distil them in a Retort in Sand, and there will first come forth a white fume, which being con-densed falleth in a straight line to the bottom; next will come over a red Oyl of great *medicinal* efficasy, exceed-ing the vertues of the Salt of Nitre. For illustration of part of this process, take pure May-dew, gathered in the morning, and put it into a Glass vessel, covered with a parchment pricked full of holes, and set it in the heat of the Sun, for the space of four months; and there will store up green feces, which will fall to the bottom; the residue of the water being white and clear. Now by all this, we may conclude what manner of dissimilarity there is in the *parts* of Water."

> " With this Canon, I dwelt have seven year
> And of his science am I ne'er the near. . . .
> It slides away so fast
> It will make beggars of us all at last . . . .
> That sliding science hath made me so bare . . . .
> And of my swink (labor) yet bleared is mine eye.
> Lo ! what advantage is to multiply . . . .
> A man may lightly learn; if he have aught
> To multiply, and bring his good to naught . . . .
> For conne he letterure or conne he nonne
> Both endeth in multiplication. . . .
>
> Our lampēs burn both night and day;
> To bring about our craft if that we may;
> Seared pokette, salt-petre, and vitriol
> And divers fires made of wood and coal;
> Sal-tartar, alkali, salt preparate
> And combust matters, and coagulate;
> Clay made with horse and mannēs hair and oil
> Of tartar, alum, glass, barm, wort, argoil,
> Rosalgar, and other matters imbibing . . . .
> The four spirits and the bodies seven—
> The first spirit Quicksilver called is;
> . . . . And the fourth Brimstone."
> *Chaucer, in " The Canons Yeoman's Tale."*

Now does the reader query how it was possible for the learned in old times to believe and do such strange things as alchemy asserts? If we turn to that judicious and profound recent author of the " Early History of Chemistry," Ferdinand Hoefer, we can see from another standpoint how delusions in those times were easily fallen into. Let us forget, he says, for a moment, the advancement chemistry has made since the fifteenth century. Suppose we were, in fancy, transported to the laboratory of one of the great old masters of " the sacred art " in the earlier centuries, and watch his operations :

" 1*st Experiment.*—Some common spring water or well water is heated in an open vessel. The water boils and changes to an aeriform body (steam), leaving, when boiled away, at the bottom of the vessel, a white earth, in the form of a powder. Conclusion : water changes into air and earth. What objection could we make to this infer-ence if we were wholly ignorant of the substances which water holds invisibly in solution, and which are, after evaporation, deposited at the bottom of the vessel ?

" 2*d Experiment.*—A piece of red-hot iron is plunged into a basin full of water, covered with a glass bell. The water diminishes in volume, and a candle being in-troduced into the bell, sets fire to the gas inside. Conclu-sion : water changes into fire. Is not this the natural conclusion which would present itself to any one who was ignorant that water is a composite body, consisting of two gases, one of which, oxygen, is absorbed by the iron, while the other, hydrogen, is ignited by contact with the flame ?

" 3*d Experiment.*—A piece of lead, or any other metal except gold, platinum, or silver, is burned (calsigned) in contact with the air. It immediately loses its primitive properties and is transformed into a powder or species of ashes or lime. The ashes, which are the product of the death of the metal (for it was early the prevailing belief that metals possessed life), are again taken and heated in a crucible, together with some grains of wheat, and the metal is seen again, rising from its ashes, and reassuming its original form and properties. Conclusion : metals are destroyed by fire and re-vivified by wheat and heat. No objection could be raised against this inference ; for the reduction of oxides by means of carbon, such as wheat,

was as little known as the phenomenon of the oxidation
of metals. It was from this power of resuscitating and re-
viving the dead—that is, calsigning metals—that grains of
wheat were made the symbols of the resurrection and of
life eternal. Wheat kernels capable of germination are
found in the wrappings of some of the oldest mummies in
Egypt.

"*4th Experiment.*—Argentiferous lead ore, appearing some like a galena, is burned in cupels made of
ivory black or pulverized bones; the lead disappears,
and at the end of the operation there remains in the
cupel a nugget of pure silver. Nothing was more natural than to conclude that the lead was transformed into
silver, then to build on this and similar facts the theory of the *transubstantiation* of metals, a theory which,
later on, led for ages a searching for the philosopher's
stone.

"*5th Experiment.*—A strong acid is poured upon copper—for sulphuric, nitric, and hydrochloric acid were early
known—the metal is instantly acted upon, and in process
of time disappears, or rather is transformed into a green
transparent liquid. Then a thin plate of iron is plunged
into this liquid and the copper is seen to reappear in its
ordinary aspect, while the iron in its turn is dissolved,
and there is a green liquid. What more natural than to
conclude that iron is transformed into copper? If, instead of the copper, a solution of lead, silver, or gold had
been employed they would have held that iron was
transformed into lead, silver, or gold.

"*6th Experiment.*—Mercury, which is quicksilver, is
poured in a gentle shower on melted sulpher, and a substance is produced as black as raven's wing. This substance, when warmed in a close vessel, is voltalized
without changing, and then gradually assumes a brill-

iant red color (vermillion). Must not this curious phenomenon, which even in the present day is unable to
be explained, have struck with amazement the worship-
pers of the sacred art; the more, as in their eyes black
and red were the very symbols of light and darkness,
the good and evil principles, and the union of these
two represented in the moral order of things their transcendental God—the universe?

" *7th Experiment.*—Organic substances are heated in
a still, and from the liquids which are removed by distillation, and the essences that escape, there remains a
solid residuum. Was it not likely that results such as
these would go far to establish the theory which made
earth, air, fire, and water the four elements of the world?
The solid metals were in those days considered as living
bodies, and the gases as souls, which they sometimes
allowed to escape. Of all the ingenious inventions of the
famous Jewess, Maria, of those early times, for regulating
fusions and slow distillations, the only one that has
survived is the *Balneo Maræ*, or treatment of things by
liquefaction and *gentle heat.*"

Science generally that appeared in later times seems
to date from three discoveries, namely, that of Co-
pernicus, the effect of which expelled the astrologers
from the society of astronomers; that of Torricelli and
Pascal, of the weight of the atmosphere, the foundation
of physics; and that of Lavoisier, who, by discovering
oxygen, destroyed the theory of Stahl.

Before closing this chapter it should be stated that
the older phase of alchemy, that known to the ancients,
included astrology and was largely a speculative philosophy. It was very ancient, and at first largely religious.
Then it passed into the hands of the priest-doctors in Syria
and Egypt, and then in Greece and Rome, and so on in

all Europe until the sixteenth century, when, as we see, it degenerated into pure charlatanism. No longer was the definition of alchemy, as the pretended art of making gold, correct or adequate. It was the alchemists who first stated, though confusedly, the problems which science is still engaged in solving. Now, the full time is come to apply to all occult questions the same searching analysis to which the other myths and assumptions of the older times have been subjected. Old alchemy lent itself to the search for the philosopher's stone and the universal panacea, while it sought to point out the origin of life, and then to show the connection of different lives, or Cosmos. Professor Christlieb, at Bonn, says it taught pantheism, which now in these times stands in complete contradistinction to all sound investigation of natural science. According to the latest researches of Pasteur, which are confirmed by the French Academy of Sciences, the assumption of a "*generatio spontanea* or *acquivoca*"—i.e., that organic life should spring from inorganic matter—must hereafter be considered as scientifically incorrect. The second phase of alchemy we see exemplified in these extensive quotations from French as a mixture of ridiculous charlatanism and chemistry. While the third phase of alchemy crops out in our own time most prominently in the so-called system of homœopathy.

Retzsch Del., 1814.

THE RESULT OF ALCHEMY.

CHAPTER VII.

ANALYSIS OF HOMŒOPATHY.

> *Prosperus.*   'Tis time
> I should inform thee further.   Lend thy hand,
> And pluck my magic garment from me—So :
> Lie there my art.— *The Tempest.*
>
>   *Bernardo.*   Sit down awhile ;
> And let us once again assail your ears,
> That are so fortified against our story,
> What we two nights have seen.
>   *Horatio.*   Well, sit we down,
> And let us hear Bernardo speak of this.
>   *Hamlet.*

IT now remains for us to examine and explain the only exclusive medical " sect" of our own day. I refer to homœopathy. And we need to examine it not only fairly, but thoroughly, for it is an organized sect, which does not hesitate to assume that its particular theory and practice is the only one in accordance with true science, and teaches that the regular medical practice is wrong and injurious. Homœopathy is evidently the outcome of a long line of old-school dogmatic systems of medicine and alchemy, whose theory and practice were based more or less on *occult* and transcendental ideas. The small factions of eclectics and botanics, of faith cure, spiritualists, clairvoyants, hydropathists, and electropathists are only of yesterday, and have no lineal descent from the successive ancient schools ; while homœopathy has drifted along down the present century, very much

12

to-day, as it was at first, the avowed enemy of the regular medical profession. Homœopathy, as is well known, originated with Samuel Hahnemann, who was born at Meissen, near Dresden, in 1755. In 1797 he published his first idea in regard to this dogma. He then stated that he believed the reason that cinchona bark cured the ague is because it has the power to produce symptoms in a healthy person similar to those of ague; and this was *one* of the old-time axioms. He claimed he had made the discovery of a *law* that has no exceptions, yet no other person since that time has found it uniformly true. In 1808 he got so far as to publish his " Organon," which he continued to do up to some forty years ago, when he died in Paris.

Hahnemann called his dogma, that like cures like, " homœopathy ;" and then he nicknamed all regular physicians " allopathists," and published it to the world. He then placed "an impassable gulf between them,"—for these are his very words,—and his followers, the homœ-opathists, to this day have not failed to maintain this *antipathy* and cause it to spread among the people, until " allopathy" and "homœopathy" are familiar words in some communities. The writings of the homœopathists are full of it, their speech is not without it ; as if it were, as it really is, an irrepressible conflict.

This work, the " Organon " of Hahnemann, is still the very Koran of the homœopathists everywhere, and is invariably the first in the list of text-books at their schools to the present time. Those other various later medical systems, so much taught as systems just before Hahnemann appeared, by the famous Hoffman, Stahl, Boerhaave, and Brown, have all collapsed and passed away ; yet it is true that some thoughts they advanced still live to enrich the present symmetry of true medical culture.

That the homœopathists are an "exclusive sect" can be proved by abundant unanswerable evidence, since they themselves continue to set it forth in their books and catalogues, as well as speech. Before me are several catalogues of colleges and professors, of hospitals, and books just issued and received, as that of the New York Homœopathic Medical College, the Homœopathic Hospital and College of Philadelphia, the Chicago Homœopathic Medical College, the St. Louis Homœopathic Medical College ; the Homœopathic Pharmacy of Otis Clapp in Boston, Henpel and Breakly's Homœopathic Practice, Boerick and Tafel's Homœopathic Pharmacy, and pricelists of homœopathic books. But who ever saw a catalogue of an "Allopathic" Medical College of Philadelphia, or an Allopathic Medical College of New York, or Chicago, or of any where else on the face of the earth? Then who are the self-styled "sectarian school"? Who nicknamed the other, and continues to do so, and still represents its practice to be what it is not? The fact is, the homœopathists are the real "old-school doctors" of the present day, by lineal descent, retaining the absurd assumptions and assertions of alchemy.

"The name *homœopathy* is made from a Greek word meaning 'like' and another meaning ' affection '—that is, a like affection is *cured* by the remedy. A fanciful doctrine, which maintains that disordered actions and real diseases in the human body are to be cured by causing other and greater, or more powerful disorders of a like kind ; and this, as asserted by Hahnemann, will ' overpower and expell '—that is, cure—any disorder ; and all this is to be accomplished by exceedingly infinitesimal doses ! These pretended doses are often of inert agents, as the thousandth, millionth, or decillionth part of a drop or grain; for example, of bark, camphor, camomile, thoroughwort, 'natrum,' or table salt, charcoal, chalk, or soda, which is the actual authorized dose employed by the homœopathists."—*Dunglinson's Medical Dictionary.*

Of course that is nothing less than falsification of medicine, and whoever practises it understandingly must be a falsifier. Should this be accepted, countenanced, or condoned by the regular medical profession it would to a degree become a *similia* and a falsity. Moreover, every homœopathic physician who pretends to learning and common sense, and yet continues to teach and practice such nonsense in these days, ought to be ashamed of himself. This subject has been tested and examined thoroughly from time to time by thousands of very able and competent physicians in different countries, and it has been uniformly condemned as a fallacy or a false pretence, and a great organized scheme of quackery. It is now believed that the remedy for this is to make the "system" of homœopathy thoroughly known among the people, who unwittingly sustain it, or it would soon vanish away.*

"Scientifically conducted experiments with homœopathic *dilutions* were long ago made by Andral, and others with him, in the hospitals of Paris, and by other eminent regular physicians elsewhere, and always with negative results.† Some patients doubtless improved while these so-called medicines were being administered, but not in consequence of their administration, in the opinion of those best qualified to judge. Indeed, a large number of leading homœopathists now deny the efficacy of these imponderable doses, though *the homœopathic*

---

* If we take any notice of a quack medicine, it blows it into notoriety. If we say something against it, the cry of *persecution* is raised and the result is its temporary success. But if we examine it and analyze it, and show by an *exposé* what it is made of, what is in its dose and what is *not* in it, it dies sooner or later as surely as a body that has received a mortal wound.

† Dr. A. B. Palmer, in *North American Review*, March, 1872, p. 311.

schools and text-books still teach their use, and most of the homœopathists, though often resorting to other remedies, still give the tiny sugar pellets. As already intimated, there is not a tenet, as presented by the founder of this system, which has not been rejected by various members who are regarded as high authority in the homœopathic fraternity. The denial of the efficacy of the higher dilutions is an admission that all the reported wonderful cases of success in the past by Hahnemann and his followers were deceptive. Their treatments were claimed to have been made with their dilutions, or alleged 'potencies.' "

No, it is the *ensemble* of homœopathy, especially the ridiculous *infinitesimal dose*, and all that that carries with it or implies, that gives offence to common sense, to science, and to honest rational medicine. To-day in the city of New York is a homœopathic apothecary and bookstore, one of their foremost, from which this year, as in the past, are sent out catalogues, with price-lists, all over this great country, in which it is stated—

" Finding that our thirtieth (dilutions) gave good satisfaction to physicians, we concluded to make in like manner High Potencies ; that is, preparing them by hand, with pure alcohol, giving each potency twelve powerful shakes. We have thus carried up over two hundred and fifty different remedies to the two hundredth, and one hundred and fifty to the five hundredth dilution, then one hundred to the one thousandth potency." I know you must be ready to exclaim, Can it be possible that they expect to find customers for these great dilutions? Certainly they do, or they would not make them, or say they do. I must confess that both the making and the selling seem quite incredible. Why, in real fact and truth, it would take all Lake Erie, Lake Michigan, and

Lake Superior to hold the whole of that *one drop* when so diluted! Yes, it would require a demijohn larger than the moon or the earth to contain all of that homœopathic drop or grain when so diluted or attenuated.

Am I too severe when I denounce homœopathy as being a sham, a falsification, an organized quackery? Here I appeal to all fair-minded and eminent men in science and learning if I have overstated the facts, or exaggerated one iota, as based on the homœopathist's own asserted and claimed dilutions and triturations. In fact, I am prepared to state that there is no appreciable medicine whatever in the homœopathic fifth or sixth potency, which are of their *lowest* class of dilutions. Then if there is no medicine in their dose, homœopathy is false. This is self-evident. All its testimony from beginning to end is false, or a fallacy. Homœopathy has been abandoned, and real medicine has been given by the homœopathist *in every case* where the necessity required and *relief or cure was obtained from medicine.*

Different writers at different times have made various calculations with the view of giving as clear and definite statements as possible of the "infinitesimal doses," which are the very core and heart of the system of homœopathy, and in so doing they usually follow the precise successive dilutions up to the thirtieth attenuation or potency, as Hahnemann taught. The author does not propose so much, but only to dissect thoroughly, and demonstrate clearly, the least dilutions termed " low potency." Dr. Hempel, a prominent homœopathic authority, places the different dilutions they use into four grades of classification.

First class—the lower, from the first to the sixth potency.

Second class—the middle, from the sixth to the thirtieth potency.

Third class—the higher, from the thirtieth to the two hundredth potency.

Fourth class—the highest, from the two hundredth to the one thousandth potency.

It is those of the first class only that we will try to measure and comprehend, for even those of their fifth potency, which are of the larger doses peculiar to homœopathy, are so small and attenuated that with the most powerful microscope yet invented we cannot see any medicine ; no one can find in them by chemical tests the least trace of medicine. This statement we are prepared to prove. Of course not only the lower dilutions, but also the higher, are claimed or employed by the homœopathists, for when the utter nonsense of the " infinitesimal " is abandoned homœopathy is gone. " The higher the infinitesimal the purer the homœopathy." As to the " similia " dogma, that is of little consequence either way, excepting as it is put forward as a figure-head for the invisible " system." It is not a universal law, as they had supposed, and they now know it is not ; for many of their leading members already have abandoned that claim. Any honest and earnest physician may choose his particular remedy for a particular case by that rule, or any other rule, if he finds it of any advantage to do so ; but that is no reason why he should cut himself off from the regular medical profession, and join some small exclusive sect, and so oppose, by so much as his position allows him, all unity and progress throughout the medical world.

After all, the homœopathist will say, " Our *theory* ought to be understood. We do not depend on the large size of the dose." But what is the sense of talking about "the dose" when there is not a particle of medicine to make a dose out of? Homœopathic dilutions utterly attenuate it (the drop) until it disappears like a vanished cloud. There is nothing there but a grain of sugar or a drop of alcohol. But the homœopathist replies, "We need to have present the *similia* of a nonentity of disease, manifested by a symptom, to *attract* the nonentity of the potentized *dose* in order to get a homœopathic cure." Hahnemann says, and they still claim, that there is no other cure but the homœopathic. That is, there must be "a spirit-like aberration in the patient," producing certain symptoms called disease, which requires a *similia dose* of a nonentity to neutralize it and *drive* it from the system ; because *then* the pellet, a particle of sugar, becomes more powerful than the disease! Conclusion : this is a fair sample of homœopathic logic, theory, and practice.

Now let us see how the dose is made. We must first bring out the process into broad daylight. Let us examine it in the tiny vials on the table *secundem artem homœopathicum*, then work it out on the blackboard, and set the boys and girls at work " proving " the so-called " doses." Almost any little American girl or boy over ten years of age, if asked to try it, can explain, " prove," and show up this system of homœopathic fallacy, or falsification and sham, which hundreds of otherwise well informed people do believe and do not examine.

Hempel, the eminent and recent homœopathic writer, says: " In order to obtain good homœopathic preparations follow Hahnemann's rules as closely as may be possible and convenient."

> " The insane root,
> That takes the reason prisoner.
>
> .   .   .   .

And oftentimes to win us to our harm
The instruments of darkness tell us truths,
Win us with honest trifles, to betray us
In deepest consequences."—*Macbeth.*

The origin and authority of "the infinitesimal dose,"
and how it is made, are found in the following quotations from Hahnemann's "Organon."

" If two drops of a mixture of equal parts of alcohol
and the recent juice of any medicinal plant be diluted
with ninety-eight drops of alcohol, in a vial capable of
containing one hundred and thirty drops (for the convenience of shaking), and the whole be *twice* shaken together, the medicine becomes *exalted* in energy to the
first development of power, or, as it may be denominated,
the first *potence.** Then the process is to be continued
through twenty-nine additional vials, each of equal ca-
pacity with the first, and each containing ninety-nine
drops of alcohol, so that every successive vial after the
first, being furnished with one drop from the vial or dilution immediately preceding (which has just been shaken),
is, in its turn, to be shaken twice, remembering to number the dilution of each vial upon the cork as the operation proceeds. These manipulations are to be conducted
with care through all the vials, from the first up to the
thirtieth, or decillionth development of power, which is
the one in most general use."

" In addition to this dilution, it must be remembered
that the power of homœopathic medicine is augmented
(potentiated) by friction and shakings at each successive
division and comminution. This development of powers, unknown before my time, is so great, that in later
years convincing experience has led me to make use of

* Hahnemann's "Organon," p. 200.

two succussions (shakes) after each successive dilution, where formerly I employed ten." *

According to Hahnemann, then, we are to take thirty, two-drachm vials with corks, and stand them along in a row on a counter or table before us. Into the first vial must be put *one drop* of the juice, or one grain of the prepared powder, that is to be diluted. This is all the medicine to be used of any one kind in all the potencies from the first to the thirtieth. But we shall follow the process only to the sixth dilution or potency; I am sure this will be far enough to be completely convincing. To this end, then, we need only to take six two-drachm vials. Into the first we put exactly one drop. Then we are to dilute this drop by adding ninety-nine drops of alcohol, shake it twice, and cork it. Then one drop of this dilution is to be put into the second vial, adding ninety-nine drops of alcohol, shake it twice, then cork and mark it two. Then the third vial is to be treated like the last; and so on with the fourth, fifth, and the sixth. Now a drop of this sixth dilution is a dose, or rather a fraction of a drop of *it* is used to medicate small pellets, or is put into water, which again is divided into teaspoonfuls. How very simple! Does the reader suspect any fallacy here? Can you tell, or even comprehend at once, how much, or rather how little, medicine there is in such a dose? Look out, *for there is fallacy in this deal!*

That is, in the first dilution of one drop, or one grain, of the medicine, one drop contains the $\frac{1}{100}$ of the original drop. Then one drop of this dilution added to the next vial forms the second dilution, so that one drop of this contains the $\frac{1}{100}$ part of $\frac{1}{100}$, which is $\frac{1}{10,000}$ part of a drop. The third dilution contains $\frac{1}{100}$ part of the second, so that

---

* Hahnemann's " Organon," page 222.

one drop of it contains $\frac{1}{1,000,000}$ part of the medicine drop. The fourth dilution contains the $\frac{1}{100}$ part of the third, so that one drop of it contains $\frac{1}{100,000,000}$ of the drop of medicine. The fifth dilution contains the $\frac{1}{100}$ part of the fourth, so that one drop of it contains $\frac{1}{10,000,000,000}$ part of the one drop of medicine. The sixth dilution contains the $\frac{1}{100}$ part of the fifth, so that one drop of the sixth contains the $\frac{1}{1,000,000,000,000}$ part of the one drop of medicine!

Can it be possible that any sane person who is honest and knows this fact would ever seriously prescribe such for the sick and suffering? In the process of these dilutions, suppose, instead of discarding ninety-nine one hundredths of each dilution in making the next higher dilution—that is, instead of taking one drop only from a vial containing one hundred drops—we take all the hundred drops and dilute each drop of them all, at the rate of one hundred times for each drop, we find we arrive at some very remarkable facts, as will be shown. Now if we proceed in this manner to the sixth dilution we find the one drop of medicine we began with dissolved in one trillion (1,000,000,000,000) drops of alcohol, which would be a bulk of liquid equal to 651,041,666 quarts, or 3,014,081 hogsheads. Then as each drop of this bulk of liquid would be a homœopathic dose we have here one trillion homœopathic doses. In other words, the one drop of drug—as tincture of camomile—would furnish (according to homœopathic theory and practice) about five hundred doses to each and every human being on the face of the earth. Of course it is needless to attract the attention of any intelligent person to the utter fallacy of any such practice. But think of their pretending to carry this diluting on to the tenth, twelfth, twentieth, or thirtieth, or even to the hundredth dilutions, and actually pre-scribing drop doses of such for the sick!

Whereas in fact it is well known by all our apothe-caries in Boston, and by the patients of the homœopathists who testify, that they resort to medicated suppositories and parvules in *full strength*, the same, only more fre-quently, than the regular physicians do. Indeed, the homœopathic apothecary in Boston is now advertising for sale these medicated suppositories of full medicinal strength—as opium, one or two grains ; also belladonna, one grain, or half to one-quarter grain ; quinine, two grains ; creosote, five drops ; carbolic acid, one grain ; morphine, one-quarter grain ; iodoform, one grain.

There is reason to believe that the homœopathic people have not clearly understood or comprehended even that very low potency called the third, or they certainly would not be deceived by it, though they are directed by the homœopathists to make much use of it in domes-tic practice. This potency or dilution is used by all shades of homœopathic practitioners, pure or partial, of high or of low potencies. What is true of this is true also of all their higher potencies. Therefore let us ex-amine this so as to clearly understand it. The third potency, the same as all the other potencies, is directed to be made in very small vials, and *this is a blinder* to be-gin with. For example, one drop of some herb *tincture* is first put into a little vial, and to this is added ninety-nine drops of strong alcohol, which makes the first dilu-tion. Then they take one drop from this one hundredth dilution and add to it one hundred drops of alcohol, then make certain shakings, when this becomes the second potency. Next they take one drop of this ten thousandth dilution and put it into another small vial and add one hundred drops of alcohol, as before, together with a cer-

tain number of shakes, and no more, which constitutes
the third dilution, called potency, according to the rule
of Hahnemann and the modern homœopathists of all
shades. This they do not deny. But do you notice that
each time they throw away ninety-nine drops out of
every one hundred, that is, $\frac{99}{100}$ of it, that you hear no more
about? It would be the same if the dilutions were re-
peated to the thirtieth or the three hundredth potency,
you would hear only of the one hundred drops.

Now the question is, are not the people deceived by
this kind of manipulation? Do they realize that the first
one drop of tincture is diluted in the third potency so as
to be one million drops, or at the rate of two drops to a
barrel of alcohol? But is this so? If we multiply the
first drop by one hundred, and then multiply this one
hundred by one hundred, we get the second potency,
which is ten thousand drops. If this is multiplied by
one hundred we get the third potency, which is one
million drops, when fairly and fully carried out so that
you can see it.

Then again if we reduce this by dividing the one million by sixty, because sixty drops make one drachm or
teaspoonful, we see we have sixteen thousand six hundred and sixty-six drachms. If this is divided by eight,
because eight drachms make one ounce, we see there are
two thousand and eighty-three ounces. Then if we divide this by sixteen, because sixteen ounces make one
pint or pound, we find we have one hundred and thirty
pints. Then if we divide this by two, because two pints
make one quart, we have sixty-five quarts. Divide this
by four, because four quarts make one gallon, and we see
we have sixteen gallons, or a half barrel of alcohol, in
which, or at which rate, the original one drop of tincture

is diluted, and which is termed the third potency. The dose of this is one drop, or a fraction of a drop !

Now suppose the whole of this dilution, called the third potency, were put into four large pails holding some five gallons each, or into a large tub holding some eighteen gallons, and it were carried into a large theatre, church, or music hall that contained three thousand people, and some homœopathist should prescribe three or four drops of this third potency to be put into a tumbler of water for each person in that audience, would there be enough to dose them all? Or, rather, how much of it would be taken? Why, that audience would have to be dismissed and the house refilled a great number of times before the original *one drop* is taken up! In fact, there were one million doses, and would serve hundreds of thousands at that rate! Every reader must see that this demonstration proves their utter folly and fallacy, for there is no appreciable medicine at all in a homœopathic dose of the third potency. Then what about their sixth, tenth, twelfth, twentieth, and thirtieth potency? They are all *similars*, for there is no real medicine in any of them. It is a process that results in nonsense and fallacy.

From the foregoing what conclusion can reason, logic, or common sense come to in regard to so much testimony from the homœopathists that they have cured thousands *by such means !* Why, if pure vaccine should be treated to the third potency it would become inert and worthless, and the homœopathic druggists and physicians must know it. Do they ever try to cure smallpox by giving vaccine virus, third potency? Why not, if *similia* is a law, and the infinitesimal is real medicine? No, the very virus of the small-pox itself, if the scab were triturated and diluted to the third potency and

given homœopathically, would prove perfectly inert.
If neither the third, nor the sixth, tenth, thirtieth, or
three hundredth can affect the patient, because there is
no medicine in any of them, what nonsense then for them
to talk about " similia similibus curantur," or about wonderful cures, or about the spiritualized nature of diseases,
or of totality of symptoms, or of any other part of their
creed? The key-stone of their arch is dropped out when
we demonstrate that they dilute their medicine to
death, and there is nothing of it left. To be sure, there
would be real medicine, but of an unknown quantity, if,
instead of using the attenuated dilutions they claim to
employ, they should resort to a strong or *saturated* solution of some powerful drug, as tartar emetic, corrosive
sublimate, strychnine, nitroglycerine, aconitine, or atro-
pine, and employ either of these to impregnate their
larger pellets, or to use as drops; or if they should take
calomel, arsenic, antimony, or morphine, and triturate it in
their tiny, tasteless, and white powders; for they could
pass for homœopathic medicines, especially if labelled
Bell. Nux. Mercurius, etc. But this would be medical jugglery, a false pretence, and dangerous deception.
Is it possible any one could be so rash as to do this?
Then, whenever there is any medical effect, which horn
of the dilemma do they take? Or, rather, which is the
most—reckless and dishonorable?

The homœopathic " system " is therefore not only a
myth; it is a humbug and a delusion. The further we
analyze it the more we must be convinced of its sophistry. Here is an example where " a little learning is a
dangerous thing." It is necessary to hold our attention
to this subject until we see clear through " the potency,"
and can appreciate it! If it has underlying merit let us
find it and understand it, so as to believe with all our

judgment ; or else reject it.   Over forty years ago, in
1841, William Cullen Bryant, being a member of a homœopathic society of propagandists, gave them an ora-
tion, and apparently repeated or coined all the homœopathic " lingo," precisely as we hear it still.   He said
" the friends of this new practice maintain that this
minute dose " of a given potency, " although powerful
to remove disease when it coincides with it, may yet be
given without danger when it does not.   There is no
necessity, therefore, of dwelling any longer on this
point."   But we here propose to dwell somewhat longer
on this point, for there is necessity.   No doubt the homœopathists would like us to stop where Mr. Bryant did.
Yet had he known as much about the absolute inertness
of a " potency " as he knew about the beauty and power
of poetry he never would have so committed himself to
this ridiculous infinitesimal scheme.

But we do not yet fairly estimate the dose of the sixth
potency.   We are not accustomed to think or speak of a
thousand-billionth part of a drop.   We do not readily
comprehend it.   Then let us place it on the blackboard,
not as a proposition in algebra, but in simple arithmetic,
so that the girls and boys may see if we calculate correctly.   To find how dilute the sixth homœopathic dilution, called " potency," is, we have only to multiply the
first one hundred drops by one hundred, and so on to
the sixth ; thus :—

100 drops, 1st dilution.
100
———————

10,000 drops, 2d dilution.
100
———————

1,000,000 drops, 3d dilution.
100
———————

100,000,000 drops, 4th dilution.
100
———————

10,000,000,000 drops, 5th dilution.
100
———————

60 drops make one
drachm :   60)1,000,000,000,000 drops, 6th dilution.
———————

8 drachms one
ounce :   8)16,666,666,666 drachms.
———————

16 ounces one
pint :   16)2,083,333,333 ounces.
16   2)1,302,083,333 pints.
———   ———————
48   4)651,041,666 quarts.
48   ———————
———   120)162,760,416 gallons.
033   120 (1,356,336 large hogsheads
32   ———
———   427
133   360
128   ———
———   676
53   600
48   ———
———   760
53   720
48   ———
———   404
53   360
48   ———
———   441
53   360
48   ———
———   816
5.   720
                     ———
                     96.

Thus we find the sixth dilution or potency the *one drop* is diluted in, and at the rate of, is one million three hundred fifty-six thousand three hundred thirty-six large hogsheads of water or alcohol! The above figures prove this to be so. Then how utterly absurd! Moreover, just think of a homœopathist prescribing for the sick and suffering their tenth, twelfth, twentieth, or thirtieth dilutions, which they term "potency," as if powerful! It can be only a powerful fallacy.

Think of 1,356,336 large hogsheads of fluid with the one homœopathic *drop* lost in it, and then pretend that a drop of this great dilution is a dose of medicine for a seriously sick person! Suppose this one-drop solution of over a million hogsheads in bulk was contained in a great reservoir, whose depth and diameter to the surface of the liquid are equal, and then the perpendicular sides of the reservoir are still higher by many feet, and you should fall into it; would you not see yourself lost? Suppose ninety-nine men and women also fell in with you, would not all of you be drowned if no one came to your rescue? But a professor in the homœopathic school contended with me lately that the drop diluted in the fifth and sixth potency is not lost, but only "potentized." It is my belief, that if he should fall into such a *homœopathic reservoir* he would find himself sufficiently potentized to call for help. I am sure we would lend a helping hand to rescue many a homœopathic teacher or practitioner out of this diluted drop, hoping that the great dilution, so taken, would cure him of the nonsense of the sixth potency and of all homœopathy.

Again, if you should politely ask for a tumbler of milk because you felt sick and faint from long want of food and from fatigue, and some one should get it, but first pour it into a large cistern of water and stir it well, telling you that that was done to potentize it—that is, make it stronger—and then should dip out a tumblerful of the water—no, only one drop of it, and offer it to you, would you accept it? You certainly would call it a joke, or an insult. But suppose a person is seriously sick, and confidingly asks for some remedy, how then? The greatest and truest of all physicians, Jesus Christ, said, by way of illustrating a great principle, "If a son shall ask bread of any of you that is a father, will you give him a stone?"—that is, give him a false loaf? Need we do more to dissect and analyze the "infinitesimal," and show up the certainty of the lost condition of homœopathy, provided it is made thoroughly known? It is inevitable that whatever is not based on truth, but, on the contrary, has truth, humanity, philosophy, and science *all against it,* must some time collapse and pass away.

Hahnemann says, "Desirous of employing a certain rule for the development of powers in liquid medicines I have been led by experience and observation to prefer two, instead of repeated strokes of succession for each vial, since the latter method tended to potentize the medicine too highly. . . . I dissolved one grain of soda in half an ounce of water mixed with a little alcohol contained in a vial, two-thirds of which it filled; after shaking this solution uninterruptedly for half an hour it was equal in potency and efficiency to the thirtieth development of strength." " Our vital force, that spirit-like dynamics, cannot be reached nor affected except by a spirit-like (dynamic) process, resulting from the hurtful influences of hostile agencies from the outer world acting upon the healthy organism. Neither can the physician free the vital force from any of these morbid disturbances, *i.e.,* diseases, except by spirit-like (dynamic) alterative powers of the appropriate remedies acting upon our spirit-like vital force."*

Commenting on these notions Prof. Smythe † says: "According to the 'Organon,' disease consists of a disordered condition of the spirit-like force of the body, which manifests itself by certain disordered sensations or symptoms, the totality of which constitutes the disease or thing to be treated. Hahnemann thus elevates his pathology above anything material, and places it upon the same hypothetical plane with his remedies. This *transcendental pathology* is emphatically insisted upon in the 'Organon,' as can be shown by a few quotations from Wesselhoeft's translation of Hahnemann's principal work. So much importance is attached to this belief in the nonentity of disease that the homœopaths attempted to obliterate the very names of diseases as used by regular physicians; for in speaking of any particular case they would not say the patient had rheumatism, gout, or typhoid fever, but would proceed to enumerate the 'totality of symptoms,' as the proper thing to do ; and they discouraged investigation into the causes of disease. Then to effect a true homœopathic cure an artificial drug-disease must be substituted ; that is, produced by medicine which must be similar to, but stronger than, the natural disease."

Quoting still from the "Organon"—"The *solid* medicines are in the first place exalted in energy by attenuation in the form of powder, by means of trituration, in a mortar (with sugar), to the third or millionth degree. One grain of this millionth is then to be dissolved, and so brought through twenty-seven vials, by a process similar

* Hahnemann's "Organon," p. 221.
† Smythe on "Medical Heresies," p. 102.

to that employed in the case of vegetable juices, up to the thirtieth development of power."

Directions: "The best mode of administration is to make use of small globules of sugar the size of mustard seed: one of these globules, having imbibed the medicine and being introduced into some water, forms a dose, which contains about a three-hundredth part of a drop of the potentized *dilution*, for three hundred of such globules will imbibe *one drop* of alcohol ; or, placing one of these globules on the tongue, and not drinking after it." Bear in mind that Hahnemann's "Organon," giving these directions, is still the first in the list of text-books in all homœopath schools at the present time.

Hahnemann and his present followers claim, then, that medicines during the process of successive dilutions and agitations have imparted to them a spirit-like power which is not possessed by them as material agents ; and that this potentiality is increased by the number of shakings, as well as by the number of successive dilutions. Hahnemann teaches explicitly that homœopathic medicines act by their spirit-like power only, and not as physical agencies, and that diseases, being only of a spirit-like nature, cannot be reached in any other way than by a so-called dynamic or spirit-like force. One is ready to cry out, Can it be possible that such preposterous ideas, exceeding old alchemy, can continue to be seriously entertained by any intelligent rational person in these days?

Homœopathy is, then, a medical sect based on a so-called system of laws, which they claim are universal, invariable, and fixed; and these classify it as a distinct and exclusive sect, "system," and school, as the homœopaths themselves declare and desire to be known. Their first great law is, similia similibus curantur; or, in plain

English, "like cures like." This being their first great
law, the second is the "infinitesimal dose," which is to be
made by certain successive dilutions,—a certain rule first
laid down by Hahnemann. These are accepted and adhered to by all homœopathic teachers and practitioners
as the two fundamental laws, upon which, when taken together, hangs all that pertains to homœopathy. In fact,
they constitute homœopathy. Of course, the "like"
selected medicine necessitates the small dose, but when
that dose is still further made "*infinitesimal*" it is peculiar
to homœopathy, and is inseparable from it. Hahnemann
did not believe or recognize an actual material pathology,
a sick or diseased physiology, as causing, attending, or
being produced by the given disease, "but only spirit-
like aberrations, which are manifested by sensations or
symptoms and actions." * "The causes of our diseases cannot be material ones." "It is impossible to
admit the existence of material morbific matter in the
blood."

"In sickness, this spirit-like self-acting (automatic)
vital force that is omnipresent in the organism is alone
primarily deranged by the spirit-like (dynamic) influence
of some morbific agency inimical to life. Only this ab-
normally modified vital force can excite morbid sensations in the organism, and determine the abnormal func-
tional activity which we call disease. This force, itself
invisible, becomes perceptible only by its effects, ...
that is, by symptoms of disease in the visible material
organism."

It may be said that the homœopathic teachers and
practitioners do not now endorse or employ the higher
potencies. To show that they do, I here again quote,
not from a rare book, but from the "Homœopathic Do-

* Hahnemann's "Organon," p. 23; also p. 68.

mestic Physician," by J. H. Pulte, M.D., of Cincinnati, Ohio ; twelfth edition :—

" *Notice* to homœopathic pharmaceutists. It will be seen, by reference to the list of medicines, that one remedy only appears indicated there, in two different potencies, viz., *Belladonna* (low potency) and belladonna [∞] the two hundredth potency. This distinction should be strictly adhered to, as in the body of the work special reference is made to this fact. Yet another wish is here expressed, which, it is hoped, may be realized by the pharmaceutists who now serve the wants of the increasing multitudes of homœopathic patrons in this country." It is this :—

" Every pharmaceutist should possess himself of a complete set of reliable *higher* potencies, so as to be able to supply the wants of the people when asked for. We can safely recommend *for this purpose* the following from among those named in the list of medicines, as designed to be contained in the cases or boxes accompanying this book."

" List of medicines to be furnished in the *higher potencies*, if so desired : *Arsenic, Calcaria carb.* (carbonate of lime *), *Cantharis.* (Spanish fly), *Carbo-veg* (wood charcoal), *Causticum* (potash), *Cina* (worm-seed), *Coffea.* (coffee berries), *Colocynthis, Conium* (hemlock), *Cuprum* (copper), *Graphites* (black lead), *Ignatia* (a bean), *Jalapa, Kali hydriod.* (hydriodate of potash), *Lachesis* (the poison of snakes), *Lycopod.* (club moss), *Mercur. sublima. corros.* (corrosive sublimate), *Natrum mur.* (table salt), *Petrol., Phosphor., Platina, Sepia* (the inky juice of the cuttle fish), *Silicea* (silicious earth), *Stannum* (tin), *Sulphur.* These are the most important of those which show better results in the higher than in the lower potencies. All the

---

* These parentheses, for explanation, are added by the author.

others may be furnished, as heretofore, in the third and sixth potency." Hahnemann and homœopathists still divide all their medicines for further dilution, strong and weak *alike*, into *one drop*, or one grain each, whether it be arsenic or coffee ; corrosive sublimate, calomel, or clam shell ; lachesis or lycopodium ; belladonna or camomile ; table salt or the stings of bees. How admirable !

" It will be seen that in the addressed *notice* to homœopathic pharmaceutists on the foregoing page," says Dr. Pulte, " they were expected to furnish such remedies as I named, in the higher potency, if desired by the people. As a great many have their medicine-chests yet filled with medicines of a low potency it was not deemed desirable to designate this change in the 'list of medicines' by adding the letters ᶜᶜ· to the names of such remedies as we would like to see used in that high degree. Each one can exercise his own judgment on the subject, and if he wants to follow our advice he can ask of his pharmaceutist for these remedies in the higher potency, and he will find them very effective. As to our views on the preference of the higher or the lower attenuations, we would remark that we consider homœopathically legitimate and practically useful all potencies, from the mother tincture and first trituration up to the *highest* dilution." The ᶜᶜ· implies the two hundredth centesimal dilution and high potency. But there is no medicine there, hence all the past testimony of its great usefulness is a complete fallacy.

Yet on page 554,* under the head of Convulsions, Spasms, or Fits, in prescribing for convulsions from *teething* he says: " Give *Coffea* and *Bella.* ᶜᶜ· in alternation, every ten minutes, two globules ; these are gener-

* " Homœopathic Domestic Physician," by J. H. Pulte, M.D., p. 554, 12th ed.

ally sufficient to allay an attack of this kind." Again, under the head of Softening of the Stomach, on page 571, he says : " *Secale corn.* is the specific in all its stages. Other remedies, such as *Tartar emet., Bryonia, Hellebore, Arsenic, Phosphor.,* etc., may arrest its progress ; but none has such a complete control over it as *Secale,* when it is in the thirtieth potency—ten pellets or globules dissolved in water, of which a teaspoonful should be given every two or three hours." But there certainly is not a particle of medicine there, and every intelligent homœopathist ought to know it.

What is Homœopathic Practice?

"But one law of cure has ever been established, and that is *similia similibus curantur.** Let us briefly examine this law, which we may do by answering the question, What is Homœopathy ? The disturbance created in the human system by morbific causes produces in the organism a mass of symptoms which represent the actual malady or disease. The object of the medicine, then, is to annihilate these symptoms, for in doing which the internal change on which the disease is founded is also removed." Suppose we try to smooth the waves of the sea, will that overpower the gale and stop the wind from blowing?

"There is a law by which we are to be guided in the removal of disease, as immutable and unchangeable as the laws which govern the heavenly bodies. No cure can be performed unless in obedience with this law of homœopathy, from which if we depart we embarrass instead of aiding nature in its operations. A medicine taken

* Guernsey's " Homœopathic Practice," pp. 564–566, 8th edition, pub. by Otis Clapp, Boston.

into the healthy system produces a certain disturbance, which gives rise to a peculiar class of symptoms. In other words, a disease is produced by artificial means." But will it be something like a typhoid fever, an inflammatory, nervous, or bilious fever, or diabetis, pericarditis, gastritis, pneumonia, pleurisy, a diphtheria, or consumption?—any disease, or assemblage of symptoms, to order? "Now where a disease is produced by other causes, with symptoms similar in every respect to those produced by the drug, we of course conclude that there is a similar internal change, in fact, a like disease to that developed by artificial means. If, then, we give this drug it is evident we produce an artificial disease. But these affections cannot exist together in the human system, for the more intense or *powerful* one will destroy the weaker. If, then, we produce with our drug an artificial affection a little more intense or powerful than the old one, that is, the disease we are treating, we of course demolish the old intruder and expel it from the system." How easy, and how simple! If it were only true! But observe the numerous *assumptions* and assertions. They equal the old alchemist in his summings-up.

"Of course the remedy must be given unmixed with any other and be allowed to produce its specific effects. The drug given must cover, or be capable of producing in health, not one symptom alone, but the entire group." Not so easy after all. For if we should take several remedies and begin by giving one only to each of a dozen people, and tell them to " prove " it, and then report to us a list of some twenty of the most prominent homœopathic symptoms produced—that is, both the grave and the trivial—how many schedules out of each set would be alike? Not two, if it were tried a hundred times over. Then, also, what about that greater intensity of disease *produced* by the homœopathic dose that is to "overpower" and "expel" the old intruder, the disease? The fact is, there is no medicine there if the dose is a true homœopathic potency, of the fifth, or sixth, or tenth potency. Nor is there any *similia* if there is no medicine. Here is the rub. Then what about the wonderful evidences of cures? As a physician, I declare there is no medicine in any such homœopathic dose unless surreptitiously put there. In that case it is not honest homœopathy. All this shall be proved before we get through with this examination. Mark, the real issue and offence is the jugglery of "potency" or dose, not as to similars, for that is of little or no importance either way. It is the sham, the pretence, the quackery, that we as physicians object to.

"*The Dose*, and its repetitions.—No definite rule can be given for all cases. As a general thing in acute diseases, excepting young children, tinctures and the low potencies should be used, while in chronic cases more benefit may be derived from the higher attenuations. I am satisfied that in domestic practice the lower attenuations may be used with much *greater safety* than the higher." * But how preposterous! Yet this is in accordance with the doctrine of homœopathy—the greater the dilutions the greater "the dynamic strength." There are some that argue and try to show that the worse is the better part; others still, that wrong is right, and that black is white, and there are those who believe them. Better had it been if all such doctrine and practice had been left in the dark ages, when many of the people, learned and unlearned, believed more in alchemy than in plain facts; more in the wonderful "exaltation" of horrid medicines, made of snakes, ants, worms, and toads, to become cures for dis-

---

* See Guernsey's "Homœopathic Practice," 8th edition, pp. 13-15.

eases, and in the mysterious pretenders of "potencies" for discovering the elixir of life.

"In my own practice," says the author, "I have generally confined myself to the tinctures and the potencies ranging from the first to the twelfth dilution, more frequently giving the third, sixth, or twelfth attenuation. . . . . If the triturations are used, the size of the dose should be about as much as could be taken up on a three-cent piece, or taken up by the point of a knife. The remedy should be placed dry on the tongue, and left there to dissolve. If the tinctures are used, one drop may be put in a tumbler of water, and if for a child or a susceptible person give a teaspoonful for a dose. If too strong, producing aggravation of symptoms, use only a teaspoonful of this solution to a tumbler of water." So the people are taught that about the one sixtieth of a drop is more simple and safe than the one millionth of a drop! How very wonderful! Yet there are some people who believe it!

Homœopathists claim that each medicine must be given by itself. In the work we have last quoted from, on page 364, we find that for certain pains "arnica should be given first, followed by pulsatilla, after three or four doses. If relief is not found and the pains are severe, with restlessness, nux-v. and camomile may be given alternately; secale and cuprum may be taken alternated where the pains are violent: dose, two drops in a tumbler of water, a tablespoonful once in one or two hours, or six globules on the tongue in the same intervals." Thus several very different medicines, so-called, are put in very fast. On page 257, "for pains affecting several teeth at once :—Cham. Merc. Rhus. Staphysagra; if with face or gums swelled: Arn. Cham. Murc. Puls. Sep. Sulph. Aur. Bell. Brion"—if not all at once, yet in such proximity, if

the toothache is severe, as to remind us of the story of
the cup of tea, into which some vinegar chanced to fall
by accident, when the quaint old man said, " Never mind,
it is all the same, excepting the taste, to mix it in the
teacup or in the stomach."

For chills and fever do the homœopathists rely on the
third, sixth, or thirtieth potency? On page 51 Guernsey
says : " Ten grains of quinine may be triturated with twice
the amount of white sugar and made into ten powders,
one of which can be taken every three hours; or five
grains may be dissolved in ten spoonfuls of water, and a
tablespoonful taken in the same way, avoiding the parox-
ysms. Even if it should not return, a powder should still
be given every day for five or six days. In addition to
the above one of the following well selected—*arsen.,
nux-v., ignatia, Ipic.*—will then produce a speedy cure."
Which, the homœopathic remedies or the quinine, will
cure? Observe the sophistry. Grain doses of quinine,
or eight grains in twenty-four hours, are appreciable
doses and efficient medicine ; but where is the infinitesimal dose of *great* potency? He says, "Copavia may
be given in certain cases, a powder of the tenth, four
times a day; or, what is better, one of the copavia cap-
sules, obtained at the druggist's, given morning, noon,
and at night."

Homœopathic Provings, showing the Similia Remedy.

Pulsatilla figures very frequently as a homœopathic
remedy of great importance. What is it? They call it
the pasque flower. It is one of the anemonies or wind-
flowers, so-called because it is said its flowers do not open
until the wind blows them open. This marvellous pulsatilla belongs to the order Ranunculaceæ. The herb and

flowers are quite poisonous, acrid, and corrosive. The so-called homœopathic "*provings*" accept it as the cure for a wide range of symptoms.* " Especially adapted to female derangements, or to persons of gentle disposition, who easily laugh or weep, with disposition to catarrh, and of lymphatic constitutions ; for chronic difficulties arising from *abuse* of sulphur-water, quinine, camomile, or mercury ; also in disarranged stomach, produced by greasy food ; bad effects from fright or shame ; rheumatic or arthritic affections, nervous difficulties, etc.; measles and their secondary ailments, itching, eruptions, and ulcers ; chilblains, swelling, and heat ; restless sleep, full of anxious and frightful dreams." Also for the cure of—

" Intermittent fever, where thirst is only during the hot stage ; acute fevers ; dizziness, as if intoxicated, especially in the evening, after dinner, or when sitting ; heaviness of the head on stooping ; headache on moving the eyes ; hemicrania, sometimes with vomiting ; aching pain on stooping ; lacerating or beating headache ; also headache from overloading the stomach, or from intoxication ; pressure in the eyes, as from sand, with inflammation and corrosive tears ; swelling of the lids, stye, or dimness of sight ; inflammation of the ears ; starting and stinging pain in the ear ; swelling in and below the ear ; noise in the ear, as of water, or a crackling sound." Also for the cure of—

" Rheumatic toothache ; toothache from a cold, accompanied with pain in the ear ; drawing, gnawing toothache, coming on when eating or taking anything warm into the mouth ; nausea, with disposition to vomit ; vomiting ; sour or bitter eructations ; water-brash ; vomiting after a meal ; pain in the stomach during an inspiration, or on pressure ; pains, cutting or lacerating ; nausea after

* Guernsey's " Homœopathic Practice," 8th edition, p. 621. Pub. by Otis Clapp.

eating fat food ; abdominal spasms, especially in pregnant females; colic, aggravated by motion ; flatulent colic ; painful sensitiveness of the bowels ; constipation ; watery diarrhœa, or if consisting of a mucous, green, or slimy substance ; dysenteric diarrhœa, or after measles ; blind or bleeding hæmorrhoids ; drawing or tensive pains from the abdomen, resembling labor pains ; derangement of the menses, or if with pain, colic, nausea, vomiting, and headache." Also for the cure of—

" Catarrhal huskiness of the chest ; dry night-cough, relieved on sitting up, but aggravated on lying down ; cough with bitter, yellow, or bloody expectoration, attended with pain in the chest, or vomiting; difficulty of breathing, especially at night in bed, and in cold air ; congestion of blood to the chest ; constructive sensation, and sticking pain in the chest, especially at night ; swelling and rheumatic pains in the nape of the neck, also in the back and extremities ; drawing and jerking pains, or trembling in the limbs ; swelling of the knee with pain." What a marvellous remedy ! Scarcely an individual man or woman but has some one of these symptoms every day; then how handy! But suppose there is no medicine, no pulsatilla there ? If the pellets are of the third potency, or higher, there certainly is none ; then where is the "evidence" of provings or remedial effect except in the imagination ? . . . . Here are omitted a list of names of diseases not proper to be mentioned. But let us examine another of their " powerful " remedies.

" *Carbo-veg* " (common wood charcoal) " is the remedy for intermittent fever ; * also for rheumatic, drawing, or bruised sensation in the limbs and joints, or attacks of weakness or vertigo ; general prostration ; nettle rash ; chilblains ; granular and lymphatic swellings; herpes ;

* See Guernsey's " Homœopathic Practice," 8th edition, page 586.

itching eruptions; unhealthy ulcers; fetid smells; drowsi-
ness; sleeplessness; pain in the head; fever, with chilli-
ness, beating in the temples, lacerating pains in the bones
and limbs; last stage of typhoid fever; collapse of pulse;
morning or night sweats; heaviness in the head; beating
or pulsating headache; congestion to the head; swelling
and ulceration of the nose; bleeding at the nose; violent
coryza; stoppage of the nose; chapped lips; eruptions on
the face; looseness and aching pains in the teeth; soreness of the gums; rawness of the throat; sore throat after
the measles; loss of appetite; nausea; waterbrash; spasms
in the stomach; pain in the liver, as if bruised; stitches
and lacerating pain in the liver; swelling of the abdomen, or aching, rumbling pains, . . . constipation; burning diarrhœa; bloody stools; tenesmus; piles; ascarides
in the rectum; hoarseness; cough, dry and hard, or with
purulent expectoration or soreness in the chest; for
whooping-cough, consumption," etc., etc. Considerable
work to be done by not more than an invisible particle
of charcoal dust, if any—not even by a good teaspoonful; but by the one millionth part of a grain! Can serious thought consider this anything but trifling alike with
simple ills and dangerous sicknesses, and the credulity of
confiding friends?

"*Lachesis*, the poison of the lance-headed viper,* for
the abuse of mercury or the effects of intoxications;
also in severe and protracted fevers; nervous disorders;
for erysipelas; also in various diseases where there is a
sinking of the vital powers or tendency to suppuration
or mortification; great weakness of body; paralysis;
hemiplegia; convulsions; rigidity of the limbs; ulcers,
with fetid discharges; melancholy; insanity, with tendency to suspicion or malice, or for fulness and dulness

* See Guernsey's "Homœopathic Practice," page 609.

of the head; weakness of the memory; difficulty to think; apoplectic fits, with distortion of face ; dizziness ; daily headaches, with languor; loss of appetite ; headache from heat of the sun, or after a cold, or daily in the morning; pulsating in the head; twinkling before the eyes ; inflammation of the eyes or eyelids, or ulcers on the cornea ; soreness in the ears or nose, with discharge of yellow matter and blood; lockjaw; toothache ; swelling of the gums; salivation ; paralysis of the tongue after apoplexy; difficulty of speech ; soreness or gangrene of the tongue ; burning pain in the throat, with redness, ulceration, etc.; nausea ; vomiting, especially of drunkards . . . .; swelling and induration of the glands; distended and hard abdomen ; inflammation or abscess of the liver; chronic constipation ; alternate constipation and diarrhœa ; swelling and pains in the larynx; cough, with ulcers in the throat; affections of the heart," etc.

Wonder if this venom is taken from the fang of the copper-head, the rattlesnake, or the adder? They all are lance-headed. The old alchemists always specified what kind and condition of snake was to be chosen for "exaltation" into medicine. (See the prescriptions in Alchemy.) When we think of the homœopathist in these days prescribing apis, or sting of bees ; or lachesis, the poison of a South American snake, we are reminded of the old story of Meg Merrilies. When she offered the honest dominie some of her devil's broth, and he hesitated to take it, she said, "Gape, sinner, and swallow it."

In 1877 Dr. Wild, Vice-president of the British Homœopathic Society, wrote to Dr. B. W. Richardson as follows:* "First, it is true that the views of Hahnemann are often extreme and incorrect. Second, that Hippocrates was right when he said that some diseases can be treated

* *London Lancet,* 1877.

best by contraries and some by similars ; therefore it is unwise and incorrect to assume the title of homœopathist. Third, that although many believe the action of the infinitesimal in nature can be demonstrated, its use in medicine is practically, by a large number in this country, all but abandoned."

Therefore Hahnemann *did not discover* the law of *similia*, as he claimed, for it was on record two thousand years before his time, in connection with other such rules or sayings. Nor was he the first to " test " medicines on the healthy. That had been tried and repeated ages before. Haller, the great physiologist, also made much of " testing " the action of medicines in healthy persons to find out their physiological effects ; but as many medicines act differently in disease from what they do in the healthy, and again differently in very small doses from what they do in full doses, but little was gained by these experiments. Hahnemann was well aware of these extensive " testings " by Haller, for he quotes him in his work ; yet in his system of homœopathy he claims to be the first to so test drugs, and then goes on beyond small doses and builds his theory and practice on the " infinitesimal " doses—that is, not medicinal doses, but assumed spirit-like or " potentized " doses—to suit his assumed pathology of " spirit-like aberration " in the human body, which may be fairly termed the third phase of alchemy.

I will here quote the following sensible remarks of a doubtful homœopath, found in the *Homœopathic Times* of January, 1878.

" In my judgment, we have sufficient evidence to warrant us in the belief that many diseases are removed when drugs are administered, which, if taken by a person in health, would produce certain morbid conditions, in contradistinction to the *host of symptoms* gathered from

the patient ' by provings' which are as likely to be im-
aginary as real, and the result of fancy as from medicine ; for we all know that no two persons will give us
the same account of their sensations and sufferings, even
though they may be the subjects of the same identical
disease in every particular, so far as we can determine.
So that any system of medication that proposes to use
drugs which in their minute details resemble the endless
phases of diseased action, lays down a proposition utterly repugnant to common sense; for the man who is
ready to avow that he understands the complications of
disease, and can interpret all its mysterious development,
so that he could apply the most attenuated *atom* to a
remote organ, passing as it must through the complicated mechanism of the human body, which in itself is
the epitome of the universe, should be declared by all
men of thought either a knave or a fool."

The above writer expresses our opinion very nearly,
and having this opinion and seeing the evil consequences
of the homœopathic faction, I am here trying to explain
as best I can the real workings of the homœopathic " system." He says, too, that he believes that many diseases
are cured when drugs are administered , so do I; but the
homœopaths only suppose or pretend to give medicine
when they prescribe those globules of the third or higher
potency, for in the third there is nothing there but the
one hundredth or three hundredth part of a drop *of the
one millionth part* of one drop of the tincture.

The real efficacy of any homœopathic medicine prepared by their rule, from the third to the thirtieth potency,
and administered in the manner they direct, cannot be sus-
tained on any rational principle. Those who have really
made intelligent and thorough examinations of these so-
called medicines, doses, and provings must and do ad-

mit this.   Those who believe in homœopathy, says Professor Palmer,* " do so on the evidence of the statements
of others, or from the supposed effects which have been
observed during the treatment of cases of disease by
the homœopath.   But it is well known that very many,
if not all, professed homœopathic practitioners often give
actual medicines in appreciable doses, and not at all
according to their exclusive ' universal law of cure.'
Thus the real relief afforded and cure effected by them
may not have been in the least due to the homœopathic
doses,"—though the people are left to suppose it to be
the effect of homœopathy.

" The system of homœopathy has been urged for the
past eighty years, and up to the present time three
generations of medical men have come and gone, while
not one educated physician out of a hundred the world
over has expressed views favorable to this exclusive
dogma and sect of the homœopathists.   On the contrary, the great body of scientific medical men everywhere, and almost all those of acknowledged eminence
in the world of science, have denounced the Hahnemann
' system ' as the most arrant nonsense.   The people who
support it are for the most part certainly ignorant of
its real character.   *Homœopathy has no position in the
world of science.*   In Europe, where it first originated, in
Germany, France, and England, the great body of those
best qualified to judge speak of this system of doctrines
now as a dream of the past, and of its practice as charlatanism and deception, while its professional adherents
are not admitted as physicians to any professional associations.   The homœopathy of to-day is no science ; not
even a dogma—only just a trade."

In the *North American Journal of Homœopathy* of

* See *North American Review*, March, 1882.

November, 1883, is an article on homœopathic practice, where *triturations* of the sixth potency are employed instead of solutions. "*Fever.*—I have repeatedly demonstrated the efficacy of Ferri phos., sixth trituration. I have given it in the hyperœmic stage of inflammation, indifferent as to the organ involved, and regardless of the cause, and always with satisfactory result. I use it thus in the prevention of traumatic fever. In the second stage of inflammation, the chief remedy is Potassium chloride. In the third stage there is a loss of the Calcium sulphate in the molecules of the tissues, and we give that salt accordingly. *Diphtheria.*—I was called to attend a child, aged six years, and found a case of fully developed diphtheria. I gave a powder of the sixth trituration of the Potass. chloride every two hours. The following day all the conditions were improved ; this remedy was continued alone for four days, and the child fully recovered on Calcis. carb."—ordinary clam shell.

" *Scarlatinal Dropsy.*—After giving Apis, Arsenicum, and Digitalis, which failed me utterly, I prescribed Sodium sulph. Twenty grains of the sixth trituration was dissolved in a gill of water, and teaspoonful doses were given every two hours. In twenty-four hours the child improved, ... and made a quick recovery with this remedy alone. *Diabetis :* a case of diabetis mellitus of six months' standing.—I treated by giving Sodium phos., sixth trituration, with entire relief of acidity in the prima viæ in two weeks, and prescribed for the next four weeks Calc. phos. sixth trituration, ten grains night and morning, with restrictions to diabetic diet." But who that understands the nature of such diseases and the effects of medicines can believe the foregoing statements ? Yet they are published as examples of homœopathic practice and efficiency.

Suppose the foregoing writer had said he put to sea on a plank one foot wide and twelve feet long, and had a very pleasant voyage, and arrived in Liverpool, England, in two weeks, safe and sound. Who that knows anything of the Atlantic Ocean, and the distance to Liverpool, and the mere raft of a plank, would believe the statement? Where should we classify such statement, with facts and truth, or with fallacy and falsehood? We must conclude, at least, that all the facts are not told exactly as they were,—that the plank was a vessel, or that a steamer took him in tow or on board; or that the putting to sea meant merely pushing off the raft from a pier to cross the water in a slip of the dock to another pier, which he called Liverpool; or that it was not a truthful statement, and that if taken literally and in sober earnest the two sets of statements are alike unreasonable and impossible. Any and every thorough chemist will tell you that there is not a particle of active medicine in any sixth homœopathic "dilution" or "trituration." That none can be found by the polariscope, nor by any other most delicate test known to science. No, nor in one thousand of their doses of it. If the patient should happen to take one thousand doses at once of the Sodium sulph. trituration, sixth potency (glauber salts), he would not feel the least effect from the medicine.

Finally, we must see clearly that homœopathy, an old conglomeration of sophistry and a mere dogma, still lingers among us *because it is not and has not been fully and fairly understood.* And here is my apology, not for writing up the long line of old-school dogmatics, nor for writing on homœopathy and analyzing it, but for giving it so much space; for examining it so in detail, that the common people may see and become aware of this great mysterious fallacy or quackery. It is high time that it should be known by its true name and classification.

Professional honor is smirched and etiquette becomes hypocrisy when we continue to positively or even passively condone or countenance that for which we cannot have the least respect. The "system" of homœopathy, like the "system" of Mormonism, has a mysterious fascination; both have many *proselyters*, who are industrious and successful. It is a mysterious excrescence, hydatid, or monstrosity of a dogma that ought to have died as soon as it was born—long ago. Plausible as the statements of homœopathists may sound at first to the unskilled ear, we see it will not stand in the presence of science or modern philosophy; nor will it bear a moment's serious examination in the presence of common sense. In fact and effect it is a network of fallacy from beginning to end, which is the mildest thing that can be said of it. Therefore many people need to unlearn what they had supposed refined truth in regard to it, and so rid themselves and their children from this false guide and pusillanimous escort that has been offering aid through the most dangerous passages of life!

In speaking of homœopathy, Sir James Simpson, of Edinburgh, said : "Surely common sense and common sanity both dictate to the human mind that it is utterly impossible that any such homœopathic dose, from any such conceivable ocean of dilution, medicated by a single drop, or a grain, of any drug dissolved and mixed in it, can have any possible effect upon the human body, either in health or disease. We can but conclude with Dr. Forbes, that in rejecting homœopathy we are discarding what is at once false and bad—useless to the suffering and degrading to the physician." We cannot but conclude with Simpson, Forbes, Trousso, Oliver Wendell

Holmes, and thousands of other learned physicians who have examined homœopathy that it is not *similia* that mostly outrages the rational mind, but the silly idea of the diluted *potentized dose* of nothing, that is to drive out the aberrations of occult symptoms, and so *pretend* to cure diseases.

Upon reading over the foregoing selections of old *recipes*, in the previous chapter, taken out of a hundred others less readable and less presentable,—these being, as before stated, quotations from the very antiquated books of the alchemist referred to,—one cannot fail to see the same genus, as well as similarity of proceedings, together with the extravagantly absurd assumptions and assertion in making the wonderful "potentiality," or "exalted potency," of a given medicine, alike in alchemy and in homœopathy. Evidently the many more times repeated pulverizing in mortar, with dilutions and redistillations, was an attempt, by French and others of those times, to revivify the ancient "art."

Hahnemann, evidently in the same line, took the hint and tried to outdo the alchemist by adopting the mythical theory of Stahl and then using spirits or alcohol ready made to his hand, for he was one hundred and fifty years later. He even increased the number of precise dilutions and manipulations, with certain shakes, no more, no less. The higher the potentizing process,—as the tenth, twelfth, twentieth, or thirtieth,—the greater the essential mysteriousness of the homœopathic art. Then with this mythy mystery, or "magestry," called by Hahnemann "potency," it is proposed to "overcome" and "expel" actual diseases—even by such mythical mightiness made out of nothingness, which the homœopaths pretend they prescribe for the sick.

The peculiar family likeness is strikingly seen upon

reading in unison the alchemist and the "Organon."
Hermes, French, Paracelsus, Van Helmont, Hahnemann,
and the homœopathists were and are all "similars," be-
longing to the same line of old-school doctors who first,
middle, and last were and are totally opposed to all regular scientific medicine and physicians. As in the past,
they (the homœopathists) are still doing all they can to
destroy the general confidence in physicians.

Noah Webster says, " The word alchemy is taken from
the Italian word *alchima*. It relates to the more difficult
and sublime part of chemistry, and chiefly such as relates
to the transmutation of metals into gold, and the finding
a *universal remedy* for diseases; and an alkahest, or universal solvent ; and other things now treated as ridiculous. This pretended science was much cultivated in the
sixteenth and seventeenth centuries, but is now held in
contempt."

As an example, the process of making the "homunculus" of Paracelsus, as quoted verbatim on pages 162–166,
certainly reminds us of the homœopathicus of Hahnemann—his third, sixth, tenth, twentieth, or thirtieth potency. We also may notice that the alchemist made use
of snakes, bees, ants, milipedes, snails, and human bones,
and toads. So do the homœopaths of the present day,
following still the same dogmatic school, employ medicines made of snakes,—not good fat ones, but the poisonous,—vipers (lacheses), the stings or tails of bees, spiders
of the most hateful and poisonous kind—as the tarantula
of Cuba, also the poisonous little spotted spider—and
potato bugs, as well as other disgusting things not to be
mentioned here. Their old mythical theory of the nonentity of disease, as also of remedy, are indeed similar.
I repeat, the similarity and relationship of these successive old schools are so self-evident the conclusion is quite

irresistible. As they appear not precisely identical, all
the better, for they certainly are *similia*, and for all intelligent homœopathists, and all homœopathically inclined,
who have any sense and reason left in them, this ought
to prove, by their universal law, a radical *similibus curantur*,—a complete and permanent cure for all that pertains
to homœopathy in toto. To really know what homœopathy is, is to eradicate it. This, rather, we believe to be
the great homœopathic law in practice for cure.

Finally, in the conclusion of this, all homœopathic physicians are cordially invited to lay aside their dogmatism
and pathy, and come, if qualified, and join the great body
of regular physicians, where they shall be honored ; so
that there shall be one body, one library, one faith, and
one practice.

APPENDIX.

To the law and to the testimony—testimony of the different factions of homœopathists against the "system" of homœopathy as a whole, or homœopathy condemning and rejecting itself by piece-meal and the testimony of others.

It is a question with some whether the homœopathic practice at the bedside is strictly based upon the principles they teach, and whether they prescribe their attenuations and rely upon them in serious cases. Some homœopathists say they do; for we must know there are three factions of these practitioners. One division, called the high dilutionists, prescribe the thirtieth, and so on up to the three hundredth potency. Another division of them limit themselves to the low and medium dilutions, as from the third and sixth to the tenth and twelfth. Another division prescribes the low potencies, from mother tincture to the sixth, also crude drugs and powerful alkaloids; or rather, they say they practice "either way," homœopathically or allopathically, according as the patient or family prefer. These intense divisions lead to family quarrels and sharp contentions among themselves, especially when they come together at their annual meetings; for they accuse each other of not knowing pure homœopathy, or on the other side of being mongrel. From some of these discussions we may learn that they can say harder things about the creed and nonsense of Hahnemann's homœopathy than any one outside of their ranks.

From a paper read before the Homœopathic Society in Tennessee, and published in the *American Homœopathist* of March, 1878, we quote the following: "The voluminous and unreliable *Homœopathic Materia Medica* is a great stumbling-block to the students of this new school. It seems as though the idea was to get *as many symptoms as possible* from each drug, regardless as to whether they are veritable drug-symptoms or personal symptoms peculiar to the prover, or symptoms arising from other causes. There is also the searching for new medicines among all kinds of animal and vegetable matter, some of them too foul to mention."

"Another great obstacle to the advancement of homœopathy arises from the position taken and articles published by some of its would-be leaders against pathology. It cannot be possible that they wish to lower the standard of education among the homœopaths. If we do so our downfall is certain. If we drop pathology, why not drop anatomy, physiology, and chemistry? Why not, indeed, drop every branch from their catalogue which is taught in all the allopathic colleges? They say that Hahnemann was opposed to pathology. . . . They say that pathology is materialistic. In this I agree with them fully. What are we dealing with but matter? There surely is nothing very *spiritual* in a case of cholera morbus or delirium tremens. Such an argument is too ridiculous to answer. It is to the bodily and not to the spiritual ills we are called to minister."

From the *Homœopathic Times* of January, 1878, I quote the following: "The Homœopathic Materia Medica is only entitled to the condemnation of all scholars and philosophers. For every one knows that if all the homœopathic physicians on earth could have lived and commenced 'testing' by such experimentation from the morning of creation and continued actively at it to this moment, they could not have proven one-half of the symptoms attributed to the various drugs therein contained." "From *provings*, we hear of the wonderful, exact, and minute effects of drugs stated with the greatest confidence and flippancy. They tell us about certain minute local effects of some remedy, at certain times, stated periods, or after a certain time, or they refer to a certain freak, whim, or fancy that went flitting through the brain, and being infatuated with the idea that they had found a *key-note*, they set the *spiritualized atom* at work to search out and *expel* the malady, which it does to the astonishment of all except the *doctor* who sent the pellet upon its glorious mission, because *he* was familiar with its most subtle and hidden power! With such foolish jargon the homœopathic profession is loaded down, . . and it will eventually be buried so deep beneath the popular judgment as to defy all possibility of resurrection."

"In proving a drug, as is claimed by the new school, upon the healthy organism, and demonstrating its exact nature and action, implies much more labor, and the whole thing is involved in far greater uncertainty than many suppose. The evidence we have upon this subject is so diffuse, profuse, and contradictory, that this whole system of drug-proving is not only doubted by many, but is to-day, with all the boasting of learned authors and professors, and *unlearned* doctors, a mooted question in the scientific world." "Our theory has demanded too much, and, in fact, more than men or angels can contribute or comprehend."

"We can take no part in the false and foolish doctrine of the potentization of drugs; this doctrine and delusion belongs exclusively to the province of the magician, who claims to produce the most astonishing changes in material things by the mention of peculiar words or the direction of his mysterious wand. The idea that a given substance can be indefinitely diluted, and its powers indefinitely increased by it, is an agelation that would have astonished the inhabitants of earth in the darkest and most superstitious ages of ancient Egypt. Those who can believe such an incredible wonder should not deride those in ancient times who exposed the sick in public places, or treated diseases by amulets, incantations, or charms; nor should they point the finger of scorn at the ancient men who rubbed black cats over the stomach of such as were tormented with the colic."

" To prove a drug by giving it to one or more persons and registering all their symptoms, peculiar fancies, and ideas, does not furnish reliable evidence of only the real effects of that drug; that is, evidence upon which a man is warranted to act, who holds in his hands the responsibility of human life. When we thus claim that we are familiar with all the minute actions of a drug we only assert that which is impossible and untrue, and entitles us to a place in the front ranks among the mountebanks who impose upon the credulity of mankind."

" If medicines become more active and efficient by such dilutions and shakings, then the same rule should apply to food, which under similar circumstances should become more nutritious. The principle has been tested upon milk and found to be a failure. It is now an undisputed fact that milk cannot be improved by dilution and shaking." The whole system of homœopathy hangs upon the truth or fallacy of this proposition of "potency." This is a greater question in homœopathy than similia similibus curantur, for if these dilutions make no effect as physical agents, and Hahnemann says they do not, and this spiritual or dynamic power is not developed, as we have shown it is not, then the supposed law of *similia* is necessarily a piece of folly and fallacy, because also that medicines cannot be prescribed in ordinary or physiological doses in accordance to that law without aggravating the given disease. Therefore this question needs no further discussion.

" When human wants can be met by such a 'system' of magic, when the philosopher's stone shall have been found, when the transmutation of metals can be effected, when flourish has more potency than logic, when brass takes the place of brains, and not till then, can he either by magic or muscle impart active life to inert substance; then, and not till then, can he diffuse power in remedies; then, and not till then, will the sense and logic of the world allow homœopathic *spiritualized* drugs a place in medical science."

" A few physicians of prominence in England," with a mistaken generous spirit, out of consideration for certain patients, " have suggested the propriety of meeting members of this 'sect' in consultation, notwithstanding the acknowledged utter absurdity of their professional views and practice and their well known denunciations of regular medicine. These suggestions," says Dr. Palmer, "it is predicted, will not meet with acceptance from the profession, for reasons which must be obvious to those who have followed the preceding statements respecting the homœopathic doctrine. The object of medical consultations is to benefit the patient, to secure for him, by exchange of opinions and by mutual agreement, the best possible course of treatment. It is too evident to require to be stated that there can be no agreement between a regular physician having any established professional views and a sincere homœopath. No benefit can arise to the patient from the practical disagreement which would be inevitable. The conscientious believer in the universal principle of similia similibus could not consent to the use of any remedy not selected in accordance with that law. One believing in the efficacy of infinitesimals, and in the injurious effects of medicines in natural form and sensible doses, could not consent, with any regard to the patient, to the giving of larger doses.

" If, for the purpose of securing patronage, the homœopathist pretends to a superior system in which he does not believe, and to a better practice which he does not follow, he certainly is a charlatan and a pretender, unworthy of confidence or honorable associations.

" If a regular physician, for the sake of a consultation fee, or for the purpose of obtaining popular favor, sacrifices his convictions, relinquishes measures in which he has confidence, and consents to a practice which he is sure is useless, he may be a fitting person for such consultations, but he is not an honorable member of an honorable profession. If between an honest homœopath and an equally honest regular physician there can be no agreement and co-operation in the treatment of a case, consultations between such are certainly useless. No opinions need to be expressed respecting consultations between parties, one or both of whom are insincere. Should the homœopathist abandon his system, or the regular physician embrace it, then there may be harmony and agreement ; but until then consistency and honor, no less than proper professional feeling, will forbid the unnatural alliance.

" The man who is honest and honorable and has been educated in the homœopathic doctrines, and has been brought into the homœopathic fraternity, but who has become convinced of the essential error of the system, will openly abandon it ; will no longer march in its ranks, or be called by its name.* One who rejects the homœopathic creed and is unwilling to occupy a false position will follow the example which some well known and honorable men now in the ranks of the regular profession have recently set, and by declaring his position and leaving his former associations will obtain recognition and a position which his talents and character will earn for him. These are the views most men will take. None are more positive in their declarations against the unnatural alliance than the leading authorities among the homœopathists themselves." Dr. Ray says, "The principle of similia similibus is the barrier which separates the new from the old school." Hahnemann had the sanity and the sense to say that homœopathy would ever be separated from what he termed allopathy " by an impassable gulf." And we also see that the infinitesimal dose must lock and bolt that bar-

* Dr. Palmer.

rier forever. In a recent address by Professor Oliver Wendell Holmes before the medical class of Harvard University, he said he proposed to point out some of the stepping-stones and stumbling-blocks that have helped or hindered the progress of the healing art. Finally he said: "It only remains to speak of some new methods and theories which have been the product of our century. I will briefly allude to the doctrines of Broussais and the numerical system of Louis. Already 'Broussaism' is obsolete, and almost forgotten at the present day. Louis, Andral, Comel, Trousseau, and others less run away with by a theory, killed it. Of Louis and the numerical system I could say much. But I prefer to say only that the numerical system can teach a wise and honest and diligent man much, and that it can make a foolish, dishonest, careless man a greater fool, impostor, blunderer, than nature ever intended him to be."

"Touching upon homœopathy," he said, "the only excuse I can offer for devoting any time to this subject is the fact that it has a certain hold on the community; that it has organizations which claim a better doctrine and a more effective treatment than what homœopaths are pleased to call 'the old school,' for which Hahnemann invented this nickname, also that of 'allopathy,' sometimes used by those who ought to know better. I require this excuse for introducing this subject, for in fact homœopathy has no *status* among the biological sciences. It has nothing of any practical value, so far as I know, to offer to the medical profession. It began by promising to prevent scarlet fever, which it miserably fails to do: and from that day to this it has been a romance of idle promises, slipping through the fingers like quicksilver, evaporating without residue, like ether from the palm of the hand. If any *one* of these promises had been fulfilled, if any single remedy brought forward by homœopathy,—of course in infinitesimal doses,—had proved trustworthy and efficacious, it would have been thankfully accepted by the medical profession, which welcomes every *method* of help, unless it comes with false pretences, and even then will appropriate any fraction of truth by itself which may be found to underlie the deception or delusion. If a drug is proved to be a remedy for any disease or symptom no physician objects to it that it is capable of producing similar symptoms in a healthy person. The regular *Materia Medica* has long recognized a class of remedies under the term of *alteratives*. Under this general head every so-called homœopathic remedy is, or would find its place, if any proved really valuable.

"Forty years ago I delivered and published a lecture on homœopathy and its kindred delusions. The three dogmas with which I had to deal chiefly were these:

"First. The one great doctrine which constitutes the basis of homœopathy as a *system* is expressed in the Latin aphorism 'Similia similibus curantur,' or like cures like; that is, diseases are cured by agents capable of producing symptoms resembling those found in the disease under treatment.

" Second. The second doctrine which Hahnemann professed to have established is the efficacy of medicinal substances when reduced to a wonderful degree of minuteness and dilution. The original drop or grain of powder operated on is carried successively to the millionth, billionth, trillionth, and very often much higher fractional dilution. A homœopathic dose of any of their so-called medicines is obtained by simply moistening with it one or several little globules of sugar, of which globules, Hahnemann says, it takes about two hundred to weigh a grain." The dose, then, is not the fraction of the first drop, but the two hundredth of a drop of this wonderful dilution!

" Third. The third great doctrine of Hahnemann is the following: * ' Seven-eighths at least of all *chronic diseases* are produced by the existence in the system (human body) of that infectious disorder known in the language of science by the appellation of *Psora*, but to the less refined portion of the community by the name of the *Itch*.' ' Psora is the sole, fundamental, and true *cause* that produces all the other countless *forms* of diseases, which, under the name of nervous debility, hysteria, hypochondriasis, insanity, melancholy, idiocy, madness, epilepsy, and spasms of all kinds, softening of the bones, or rickets, scoliosis and cyphosis, caries, cancer, fungus hæmatodes, gout, yellow jaundice and cyanosis, dropsy, gastralgia, epistaxis, hœmoptesis, asthma, suppuration of the lungs, megrim, deafness, cataract and amurosis, *paralysis*, loss of sense, pains of every kind, etc., which appear in pathology as so many peculiar, distinct, and independent diseases.' Can you believe that I am not imposing on your credulity when I say that I translate these words literally from Jourdan's French version of Hahnemann's ' Organon ' ? "

" Now, what has become of the first of these three dogmas ? The Encyclopædia Britannica, in its twelfth volume, published in 1881, quotes the following confession from a homœopathic journal, called the *Medical Investigator*, of the date of 1876: ' How many claiming to be homœopaths are daily disregarding the law of similia ! It is getting to be quite a rare thing to hear of a homœopathic practitioner conducting a serious case from beginning to end without using as such cathartics, sudorifics, diuretics, etc., in direct opposition to our law ; not only are these drugs used in this way, but there are some who say that they cannot be dispensed with ' ''—the testimony of homœopathy from England.

" As to the second grand principle announced by Hahnemann, there is abundant evidence that many, if not most, homœopathic practitioners make use of various remedies in their ordinary doses. I have had interesting revelations of this kind from my friend, the late Dr. 'Edward H. Clarke. But I was hardly prepared for the statement of Dr. Wilde, Vice-president of the British Homœopathic Medical Society, that, ' although many believe that the action of the infinitesimal in nature can be demonstrated, yet its

* Dr. Oliver Wendell Holmes.

use in medicine is practically, by a large number in this country (England), all but abandoned.'

" The more recent discovery of the *acarus scabiei*, the little insect, which proves to be the true cause of 'the itch,' has sufficiently disposed of the third of the homœopathic dogmas. Now, what there is left of a three-legged stool after one of its legs is pulled out and the other two are sawed half or three-quarters through, seems hardly worth sitting down upon. So far as I can take account of stock, the present assets of homœopathy consists of a pleasing designation, with sets of little phials containing minute globules, or amulets, arranged to correspond with a nomenclature, and a collection of ' provings,' which prove more about the prover than about the questions to be proved, a doctrine which slips on or off like a kid glove, according to the company in which the doctor finds himself. Why homœopathy should have so much popular currency in this country as compared with the land of its birth, or with Great Britain, is a curious question. But do not allow yourselves to believe because this new country is the favorite breeding place of Mormonism, of homœopathy, or of clairvoyance, that polygamy, though organized, is going to break up the sanctities of the American household ; or that these fancy practitioners will displace the educated, scientific, rational physician in the abiding confidence of the great American public."

"A few words with reference to Hahnemann, the inventor of homœopathy, whose vagaries still lie in the way, to be stumbled over by here and there one whose mental twist or imperfect scientific training has betrayed him into the misfortune of taking the wrong direction. Hahnemann was no ignoramus, by any means, but something a great deal worse. He was a hopeless subject of cerebral strabismus, beyond all medical, all surgical treatment. A squinting eye can be set right, but a squinting brain is too much for the art of gods or men. Whether the strabismuth involved the moral as well as the mental faculties of Hahnemann I will not stop to discuss. But when a man misrepresents all that he reads, when he borrows the most foolish things from the most foolish or erratic writers that he can possibly get hold of, then the less he has to do with books the better. In mentioning the authorities from whom Hahnemann probably borrowed his two best known dogmas I do not mean to say that what he took from an unworthy source *may* not be worthy of confidence. A rogue may have good money in his pocket, but his bills are more likely to be counterfeit than those carried by an honest man.

" It is well known what a braggart and pretender was Bombastus von Hohenheim, who called himself Paracelsus. All recognize him as the very type of the swaggering boasters who profess to work miracles by their wonderful skill and knowledge. Those who are curious will find the distinct statement of the *similia similibus* doctrine in his words, quoted in an article in a recent volume of the Encyclopædia Britannica. Whether Hahnemann

15

borrowed it from Paracelsus or not is of no very great consequence, but it is just the kind of hint a shrewd system-maker would be likely to find in just the kind of author he would be like to be searching; and its source lays it open to suspicion.

"The history of the probable origin of the *infinitesimal* medication is more interesting, and, so far as I know, has not been unearthed until I happened to strike the burrow this doctrine is likely to have come from. I chanced to be looking through the 'Ortus Medicinæ' of Van Helmont about a year ago, reading here and there as the titles of the chapters attracted me more or less. It was the Elzevir edition of 1652, and had stood on my shelves for many years. Among such titles of chapters as *Blas Humanum* and *Vis Magnetica* I noticed one with this odd-looking prefix:

"'*Butler*. I found this was the name of an individual, an Hibernian, a great personage formerly, as he represented, at the court of King James the First, of England. At the time, however, this stranger was provided with lodgings at the public expense in the jail of Vilvoorden, a town of Belgium. Here it was that Van Helmont, a very credulous, very whimsical man of genius, a believer in the Sympathetic Ointment and other nonsense of the kind, became acquainted with this distinguished stranger, who bore the family name of the Duke of Ormond. This captive wrought some wonderful cures, which Van Helmont reports.

"'The first case was that of a monk suffering from erysipelas. The Irishman dipped a certain pebble quickly into a teaspoonful of oil of almonds and instantly withdrew it. The patient took the oil, or some of it, and was cured at once. Second case: a washerwoman; complaint, hemicrania. He dipped the same pebble quickly into a teaspoonful of olive oil, gave it a lick with his tongue, and put it back into his waistcoat pocket. He poured that teaspoonful of oil into a small vessel of oil. One drop of this to be rubbed on the old woman's head. Immediate and permanent cure. Stupefied astonishment of Van Helmont; to whom the son of Erin, "My darling, if you don't get on so far that you can cure *any* disease with a single remedy you will remain a greenhorn (*in tyrocinio*) till you are a gray-beard."

"'The next patient was a nobleman; a bad case of gout, as it would seem. He was to touch the pebble every morning with the tip of his tongue, wash the lame parts with a cheap lotion, prepared in the laboratory of nature, and to be well in three weeks. "If he will make me well," says the count, "I will pay him his own price, and deposit it so that he shall be sure of it." Our friend with the pebble takes this in high dudgeon; will never help the miserable creature; does not want his money; is as good as he is. Van Helmont could not prevail on him to treat the case, and became sceptical. But not long afterwards a fat friend of his wanted to be rid of his obesity. Butler gave him a small fragment of the pebble, which he is to lick once or touch swiftly with the tip of his tongue every morning. In three weeks he was a span narrower around his thorax.

"'Van Helmont begins to have faith again, and being himself ill, as he thinks from poison, sends a flask of oil to Butler, who is still in jail, and who dips his pebble in it. One drop to be applied externally in one or more places. Entire failure of relief. Sceptical once more,—our inquiring philosopher. But, next, his wife is relieved of a dropsical swelling, and a servant-maid of an ill-cured erysipelas, and a widow of a stiff arm, all by *one* or more drops of the oil, and an abbess of loss of power in her right arm by only touching her tongue to the pebble. Then I asked Butler, says Van Helmont, "why so many women were cured át once, while I, near death and full of pains in all my joints and organs, got not the slightest relief." The Irishman gave a plausible answer, which silenced, if it did not satisfy, the learned simpleton.'

"The essence of the infinitesimal *doctrine*," says Dr. Holmes, "is in this most curious chapter of the 'Ortus Medicinæ,' which is well worth reading. Hahnemann mentions the name of Van Helmont in his 'Organon,' and I have little doubt that he borrowed his infinitesimal *doses*, smelling of remedies, and other inventions, from this chapter. Van Helmont, I may add, entirely anticipated Hahnemann in insisting upon the use of single or *un*-compounded remedies. 'I believe,' he says, 'that simple remedies, in their simplicity, are equal to the cure of all diseases.' And he adds, 'Consequently, the dispensaries, wishing to compound and correct many ingredients, lose everything, and by a hidden blasphemy, as it were, undertake to supply the divine insufficiency.'

"Where Hahnemann got his third great dogma, that *itch* is the cause of seven-eighths of all chronic and nervous diseases, I do not know; but I notice that Van Helmont has a good deal to say about those diseases from which he himself suffered for some months, and what he says may very probably have set Hahnemann thinking about it."

"Finally, of medical theories and practice the community is not a competent judge. Hippocrates said, 'Life is short, art is long, opportunity is fleeting, experience is deceptive, judgment is difficult.' If experience is deceptive for the trained practitioner, if a decisive opinion in cases of disease is difficult for him, of what value are the experiences and conclusions of wholly untrained individuals in medicine? But how can you argue with people so blindly wedded to homœopathy? They answer exactly in the same way with the *blind* man restored to sight, told of in the Gospel of Saint John. But who ever heard of a man or woman *born* blind being cured by homœopathy? Yet, as if the case is parallel, they say, 'Whether this be a quack medicine or no, I know not; one thing I know, that whereas I was —— now I am well;' and this argument, utterly fallacious as proof, will prove a sure defence to every form of quackery until the end of time."

"The underlying fact in regard to all this is that the great proportion of cases of sickness tend to get well, sooner or later, with good nursing and little or no medicine. So that homœopathy, like every other 'system' or method, the true and the false alike, has the advantage of this kind of deceptive evidence. Moreover, whenever a given case proves serious and needs some medicine, if the doctor knows anything about therapeutics and is cute, he will give some tangible remedies, or else ask for consultation with some 'regular' physician." Dr. Holmes, from whom we are here quoting, believes that the *inert*, inoffensive, and utterly useless homœo "globules" have all the virtues a name can give them, and no others. "*Not the less is homœopathy a system of false pretences.* If a man came along professing to teach history on the basis of Mother Goose; if he alleges as a scientific fact that a man did really'scratch out his eyes by jumping into a bramble bush, and did really scratch them in again by jumping into another similar bush, and takes this fact for the corner-stone of ophthalmic surgery, I do not think the Professors of Harvard University would feel themselves called upon to recognize him as a scientific and professional fellow-worker."

"We know that almost every form of medical sham and imposture can show some marvellous evidences of cures, not only apparently, but some as truly, for the imagination is a very powerful physiological agent. There are the magnetizers, the laying-on-of-hands, the faith-cure; and we cannot deny that there is such a thing as simple faith-cure, quite distinct from any specific, divine, or miraculous interposition. There are also the spiritualistics and the clairvoyants or mesmerics, as mysterious and fascinating to a certain class as the dynamatized *infinitesimals* are to the deluded homœopaths, and their pretended diagnosis or prognosis, if not their cures, are astonishingly satisfactory" to the messenger and her friends who carry the lock of hair and pay the fee. It seems probable from the numerous advertisements that more patients consult these liver-and-spleen-seeing and fortune-telling jugglers than people generally are aware of; and who can estimate the injury to the rising generation by countenancing or patronizing, or even tolerating as respectable, these various forms of false pretensions?

Hypnotism, a peculiar mental state, induced in certain persons by certain means, needs to be studied by physicians and explained to the people so far as understood, so that it may loose its fascinating wonder among certain people, and so that quacks cannot employ it as a part of their fallacious business. In this phenomenon it should be known that the objective symptoms vary with each case, and that there are recognized three principal types, the lethargic, the cataleptic, and the somnambulistic. This peculiar mental condition, known as hypnotism, is more easily produced in some persons than in others, and in some it is impossible. But staring at a small disk, held near in front of the nose for some minutes, is one method; the monotonous sensory impressions produced by light passes of the hands of the operator over the head, face, and thorax of the subject is another method; and by fixing the attention on some one object with absorbing thought can produce it; and it may occur spontaneously, regardless of the will of the subject, as during ordinary sleep in the night, from which it differs in that the latter is a natural total sleep of the sensorium, while hypnotism is only a partial sleep of a peculiar kind, somewhat resembling etherization.

It should be generally known that the hypnotic state is a subjective condition, and not an objective one. In other words, when the person goes into this state by her own effort, or by passes, or from any other means, it is not due to dominion of the will by the operator, but rather to the special yielding condition of the mind of the person, who, so to speak, has the hypnotic temperament, or diathesis. Hence the popular idea in regard to clairvoyance that one will has dominion over another resisting will is complete nonsense, and is so received in scientific circles.

Some wrongdoing is so mischievous and virulent that it not only hurts the sinner, sooner or later, it also involves and hurts often perfectly innocent persons. So the sporadic belief in and practice of mesmerism or clairvoyance, "the second-sight," spiritualism, and table-tipping, fortune-telling, faith-cure by laying on of hands or by anointing with oil and by prayer, or mind-cure, are not only a direct and immediate evil, but they, together with the baseless fabric of homœopathy, collectively and severally, are an injury not only to true medicine, but also to pure Christianity itself. Nor am I alone in this conviction. Our attention has recently been called in all sincerity and fairness, by Dr. Staples, of Connecticut, to the fact that certain parties profess, and advertise to possess and exercise, the power to cure the sick *by the prayer of faith*, and that many persons claim to have been thus restored to health. These are circumstances that cannot and ought not to be ignored by those who are interested either in the cause of medicine or Christianity. The evidence presented of the possession of such power, and the circumstances, should certainly be carefully studied, and the proofs presented weighed in the balances of science and unprejudiced enlightened reason.

He truly says " that the subject involved is of great importance, and that no intelligent person, be he Christian or infidel, will deny it. The claim set up must be either *true* or *false*. If true, if Christians do *now* possess the power claimed, it may be used not only in the relief of human suffering, but may become a powerful element, if rightly used, in the speedy triumph of Christianity over the doubts of *all* but the incorrigible wicked. But if, on the other hand, this doctrine be founded in delusion, then whatever may be the present or temporary benefits of a physical nature to comparatively few individuals, its effects cannot be otherwise than pernicious to the cause of vital Christianity in the end.

" Perhaps nothing has had a greater tendency to weaken the force of the argument derived from miracles, in favor of the divine origin of the Bible, than the *pretended* miracles of priestly superstition, and other hypocrisy. The cause of Christianity has suffered enough in past ages by false pretensions to miraculous power. We want no more shams in this respect ; but if God has endowed certain persons with the gift of miraculous *power*, all good men,—all who are interested in the triumph of Christ in the world,— would wish to know it, that the best possible use can be made of the fact in defence of divine revelation. But if the parties referred to are laboring under a delusion, however honest they may be in this respect, they certainly should be opposed and exposed by all true and good men.

" There can be no compromise. To say they mean well is not enough. *The* question is, is it true that God, through them, cures otherwise *incurable diseases* in a quick, instantaneous, and supernatural manner ? If he has wrought such cures as these, then the evidence to produce conviction in the minds of candid, intelligent men, and to put the unreasonable sceptic to silence, is available, and should be produced.

" It is not enough that some wonderful cures are effected by them ; the merely wonderful is not proof of the supernatural. All claims to the possession and exercise of supernatural or miraculous power should be put to the most crucial test, and all *honest* claimants of such *power* will readily consent to have their professions thus attested. That men have been endowed with such power in the apostolic age, that it is *possible* for them to be thus endowed *now*, must be admitted by all believers in Christianity; hence the question is not in relation to the *possible*, but to the *actual* possession of such power by those who now claim to possess and exercise it. We do not call in question the honesty of all persons concerned in this matter ; but we do most sincerely believe they are laboring in some respects under serious delusions, and that all *real* cures that have been effected are simply the result of natural causes.

" Let us take, for illustration, the case presented by Dr. Hall * in an editorial upon this question in the February number of the *Microcosm*. It is certainly the best authenticated instance of ' *faith cure* ' that I have seen. That the gentleman was *sick*, is now *well*, and that he recovered his health while under the advice and control of one claiming ' the power to cure by faith,' are admitted facts. But do these facts prove the interposition of supernatural power ? We think not. Let us see. First, it is not certain, by any means, that his trouble was *organic* disease of the heart; his difficulty might have been purely sympathetic, arising from other causes ; mistakes of this nature are common. Second, he retired from his usual avocation and was relieved from anxiety. Third, he changed his local habitation and surroundings ; and last, but not least, was inspired with the hope of recovery by

* *Leonard's Illustrated Medical Journal*, July, 1883.

faith in the person to whose control he had so completely surrendered himself. Now if his heart difficulty was not organic, and he had laid aside all care and removed to another locality, and acquired the same hopefulness of spirit by faith, say, in 'bread-pills,' would not the result have been the same ?

"Let the reader compare the above, or any other well authenticated case of cure by the claimants of apostolic power, with the real cases reported by St. Luke in the third and ninth chapters of the Acts of the Apostles, and they will find just cause to question the validity of their claim.

"There are in the cities of New York and Boston many well known persons,—blind, deformed, or crippled life long, 'from their birth,'—universally acknowledged to be beyond recovery by human power or skill. Now let one of these persons be restored to perfect health and soundness in answer to prayer, by the laying on of hands, and by the anointing of oil, or by the operation of all these combined, *then* perhaps we shall have demonstration of the supernatural and miraculous; but not till such a cure is placed beyond a doubt will the evidence be satisfactory to cautious and intelligent men and women. For all real miracles demonstrate the presence of the supernatural, and place it beyond all question or doubt.

"Take any or every miracle recorded in the Bible and the above principle will be found correct. If the statement of facts made are admitted, then miraculous interposition must also be admitted. Take for illustration the widow's son : if we admit he was dead, and that Christ restored him to life by saying, 'Young man, arise !' then a miracle is as evident as the shining of the sun at noonday. Now, we inquire, is this true of modern '*faith-cure,*' so-called ? Do they not, on the contrary, when all the facts are known and admitted, suggest doubt to the candid inquirer after truth ? Yes, in the presence of any such comparison doubt instantly crystallizes into utter unbelief in any and all so-called *faith-cures* by the supernatural.

"Take again the instance of the healing of the cripple from his birth by Peter and John. It would appear that the meeting of the parties was purely accidental ; the lame man did not go there to be healed, nor did he expect or ask to be healed ; there is not the least intimation that Peter and John went there with the intention of healing him or any one else. He was a mendicant, depending upon charity. He made an appeal to the disciples, as he did to others, for *alms.* Peter said : 'Silver and gold have I none, but such as I have give I unto thee. In the name of Jesus of Nazareth rise up and walk.' The impulse came upon Peter unexpectedly, and it may be a question which was the most surprised, the *lame* man, the apostles, or the multitude. It was in fact and effect *a miracle*, suggested and effected by *divine* inspiration and power, not by human will, wisdom, or prior arrangement. Hence it furnishes the most conclusive evidence, to all parties concerned, of the presence and power of Christ.

"There are, I think, but nine references to the manifestation of miraculous power by the disciples in the Acts of the Apostles, and these are scattered through a period of about thirty years. They are all attributed either to Peter, John, Philip, Stephen, or Paul; and in no instance is there any intimation of pre-arrangement of parties interested. A pre-arranged *miraculous cure* is not found in the entire record; nor do we find that the apostles ever made a profession to an abiding endowment to work miracles, or ever invited the lame, halt, blind, and sick to come to them to be healed; but when such cures were effected by them, they were the result, as before remarked, of immediate inspiration and impulsion by the Holy Spirit, and when not thus impelled to such action they were as weak in this respect as any other men. Divine wisdom determined when, how, and through whom these revelations of the immediate presence and power of Christ over all things should be made. It was not left to the will and wisdom of the apostles, and it is greatly to their credit that they never put forth any claim to such authority, nor pledged themselves beforehand to any such miraculous power.

"It may, however, still be urged that the Apostle James affirms that 'the prayer of faith shall save the sick.' True, and we have no wish to doubt or reject the truth of this declaration, or that it presents a precious truth that should lead and give inspiration to prayer and faith in every emergency. This we heartily and joyfully believe. But are we to accept the declaration in an unlimited sense? Did the apostle mean to say that all sick persons, in all ages and at all times, could be cured by the prayer of faith? And that this should continue at all times and in all ages? Not so; for if this be true, then man has the power of reversing the law of death, and securing immortality on earth." The real true meaning is self-evident.

He who utterly denies the miraculous denies God and His revelation, since revelation is miraculous. Seriously to raise the question whether God can perform miracles would be impious as well as absurd. The possibility of the miraculous rests upon the uninterrupted activity of a living God in the world. Its necessity arises on the one hand from the divine end and aim in regard to the world, and on the other from the disturbance introduced into its development through sin. Therefore, although miracles are supernatural they are not unnatural. Far from violating the conditions of life, of nature, or of humanity, they re-establish the life of the world which has already been deranged, and initiate the higher order of things for which the universe was created.

"If miracles are directed, as we have seen, not against the world's order, but against its disorder, why do we not find them occurring in every place where misery and death still prevail?* Sin and evil exist to this day; misery and disorder still abound in the world; why should not God continue

* Christlieb on "Modern Doubt," p. 330

miraculously to interfere for the removal of all these, and for the re-estab-
lishment of the original order?

" To this we answer, first of all : Are miracles, strictly so called, the only
means through which God counteracts sin and evil? Does He not first employ the internal influences of His Word and Spirit? And this has not
ceased as yet. Sin, it is true, still exists; but so does Christ, the Great
Physician for the maladies of the whole world; and His influence is ever
becoming more and more powerful and more extended. Are new miracles,
then, required while the old ones are still in active operation? Let us beware of an idle longing after the miraculous."

We must recognize that miracles belong to the divine *education* of the
human race. We now do not live in a miraculous period, like that of
Moses or of our Lord. We may learn from the history of miracles that,
according to the Holy Scriptures, miracles are more prominent in some
periods and less so in others, and that the former periods are always crises
in which the eyes of men are to be drawn to the fact that the kingdom of God
is on the eve of a momentous advance. The apostolic age required miracles, because it was the epoch in which the great Christian Church was
founded. Miracles, then, were in their appropriate place. In the last epoch
of the consummation of the Church, however, she may again need aid in
the final decisive struggle with the powers of darkness, the miraculous in-
terference of her risen Lord; and hence the Scriptures lead us to expect
miracles once more at that time. And as we have already admitted the
possibility of miracles at any time, so we cannot utterly deny that miracles
are impossible now; but the character, circumstance, and medium of the
miracle will help us to recognize the true from the false as easily as
Pharaoh could see the difference between those of Moses and those of the
Magicians. Miracles were evidently intended to confirm the divine mission
of those who perform them, and to add to the weight of their testimony.

Dr. E. Sanford, a homœopathic physician of Providence, R. I., was in Europe in 1857, and when he returned to this country he published a tract
through Otis Clapp, the homœopathic apothecary in Boston, in which he
testified to the popularity of Hahnemannism in England and Europe at that
date. " Homœopathy has gained a permanent hold in Europe, and numbers among its supporters a large portion of the intelligent and influential in
the communities where its practitioners are found. At the present time
there are upwards of seventy homœopathic physicians in London, all of
whom have received their medical education at the allopathic institutions.
Among the best known here are Dr. Laurie and Dr. John Epps, who adhere
closely to Hahnemann, and condemn any departure from his precepts. . . .
In regard to potency, Dr. Epps relies chiefly upon attenuations between the

twelfth and twentieth. He assured me that he never used aconite so low as the third, but often as high as the twelfth. The spread of homœopathy is advanced here, as elsewhere, by the same argument as related in Scripture— 'Once I was blind, but now I do see.' . . . . Dr. Epps is the author of a work on domestic medicine which has a large circulation."

"Every town in England of any importance has one or more homœopathic physicians. There are out of London about one hundred and sixty practitioners, exclusive of Scotland and Ireland. Homœopathy has a commanding position in Paris, the city where Hahnemann passed the last years of his life. About ninety homœopathic physicians are in practice in Paris, and some of them hold places of eminence. Homœopathy is rapidly spreading in France. The Emperor has favored its partial adoption in the army, and the Empress Eugene is among its active supporters. The Queen of Spain and the Spanish court all embrace homœopathy. The widow of Hahnemann is still living at Paris. In Italy there are twenty-five homœopathic physicians. In Vienna, Austria, there are forty. In Madrid about fifty, and in the German States many hundred of them."

"On account of Dr. Fleishman's success in treating diseases the Emperor of Austria was induced to establish an institution at Vienna for teaching homœopathy under the patronage of the government. The principal of this college is Dr. Wenub. The Hospital of the Sisters of Charity, at Vienna, was opened in 1832 for the reception of cholera patients, and for two years a half homœopathic and half allopathic treatment was pursued. In 1835 Dr. Fleishman was appointed at the head of it, and adopted an entirely homœopathic practice. There are two homœopathic hospitals in London. There are also fourteen gratuitous dispensaries, all of which are daily attended by many patients. In conclusion, it is impossible to meet any half dozen persons in England but several of them are adherents of homœopathy. At the public houses, coffee rooms, and in private houses it is the same everywhere. Homœopathy has acquired a great prominence, and it continues, in spite of opposition, to gain. in public favor, and steadily increases." Such testimony came twenty-five years ago. But how about it now, and the fact that the statue of Hahnemann has recently been pulled down in the streets of Paris?

Indeed, very different testimony now comes from over the sea. Instead of the wonderful increase there is a general acknowledgment of the failure of homœopathy in all the old countries where it first begun, and so flourished for a time. That homœopathy is decaying and dying out in England, France, and Germany is the uniform report of late, even by the foreign homœopaths themselves. The great medical profession there still have nothing to do with it, that is, homœopathic physicians have no professional recognition in the medical schools and societies, nor in any other scientific associations.

Homœopathy is not the first great popular delusion that has had a furious following, even by some otherwise intelligent people. There was the royal cure of the king's evil, or scrofula. From the time of Edward the Confessor to Queen Anne the monarchs of England were in the habit of *touching* those who were brought to them suffering with scrofula and enlarged glands, for the cure of them. At one time the practice was discontinued by the good sense of William the Third, but Queen Anne resumed it. According to the testimony of many eminent persons, none ever failed of recovering, "unless their little faith starved their merits." Several who had been blind for some weeks or months received their sight again at once upon being touched.* So widely at one period was the belief in this remedy, that in the course of a dozen years nearly one hundred thousand persons were touched by King Charles the Second. Even many physicians believed in it, and some testified to the efficacy in difficult cases.

Then there was, about two hundred years ago, the wonderment and extensive belief in the "weapon ointment" throughout England and Europe. Its healing power was testified to by men and women of credit. The weapon ointment, called also the *Unguentum Armarium*, was made after different methods or formulas, but of some substances addressed to the imagination rather than to the wound, for it was said to contain certain portions of a human mummy, or of human blood, and of moss from the skull of a thief or pirate who had been hung in chains. But in its use it was not to be applied to the wound, which was to be simply washed and bandaged and kept at rest, while the ointment was applied to the weapon itself with which the wound was inflicted, or to one like it. How strenuously this was believed in and practised !

Then more recently, and in our own country, there came a monstrous fallacy which commanded a widespread following, not only in Connecticut, where it originated, but also in other states, in England, and in Europe. I refer, of course, to the *ism* of Dr. Perkins, which had the semblance of magnetic influence. He employed simply two pieces of metal, shaped like small marlin-spikes, one of iron the other of brass, about five inches long and two inches in diameter at their large end, tapering off to a point at the other end. They were to be held, one in each hand of the operator, over certain places on the head, points downwards, but without touching, and in certain other places and relation to each other, and for a certain time. Dr. Perkins called these his " Metallic Tractors." He claimed that they could draw pain out of any part, and could put strength in any part, and do other wonders, and he sold them for twenty-five dollars a pair, which he succeeded in doing very rapidly and generally in our cities and country, and even more still in England, France, and Holland, until he was very rich. But he was expelled from the Connecticut Medical Society, and has since been execrated as an

* *Edinburgh Med. and Surg. Jour.*, vol. iii., p. 103.

arrant quack; and his tractors are entirely dead and gone, except as they are still now and then referred to as an example of fallacy and folly.

Homœopathy rests on the same foundation as those metallic tractors, and no more; and that is, *assertion* and *semblance*, or *similia*. The one has had its day, and is gone; the other cannot but follow in due time. But then Perkinism not being the first hobby-horse that has been ridden furiously, homœopathy is probably not the last, though the smallest in the present century.

Electropathy was flourishing here and everywhere twenty-five years ago. Among others active in propagating this *pathy* was a so-called Professor Page, who gave grandiloquent lectures before classes to teach them the "system." He showed that "in the practice of electropathy, as in the practice of every other 'system,' there must be first made a correct diagnosis by the examination of the symptoms in the case. The electric state or condition of the patient must be carefully determined. Does the patient receive a proper supply of this element of life? Are each of the avenues through which it is received in a healthy state? Everybody needed this examination to see whether they are sick or well. Is there any local obstruction to prevent its action in any part of the organism? He quotes Professor Faraday as asserting, "that from actual experiment, he finds that a single drop of cold water is possessed of electricity enough to charge eight hundred thousand Leyden jars of the usual size. What a surprising quantity does the system then receive in drinking a full glass of water!" Almost as prodigious as homœopathy.

At that time it had been the custom, in using the electro-magnetic battery or the magneto-electric crank machines, for the patient to hold the "tin handle" electrodes, one in each hand, or sometimes applying one to some part of the body or limbs. But "Doctor" Page improved on this, by placing only one electrode (the positive) in the patient's hand, while the other, a copper or zinc plate, was placed at the feet, or else was sat upon next to the skin. Another "electropathic" method was to have the patient undress and then sit upon the copper plate, as the negative electrode, while the operator held in his own hand the other electrode, and then applying his other hand or fingers as an electrode to the surface of the patient, and so passing it entirely over the whole body and limbs at a séance. He claimed that by thus passing the current through his healthy body to the patient the electricity was more vitalizing and curative. Of course this method made it necessary for every patient, male and female, to disrobe, in order to be done all over, which he also termed general faradization.

To each of the graduates of his classes he issued an engraved diploma as a master in electropathy; and as he travelled from town to town and from state to state it is believed, as he claimed, that he gave these to hundreds of physicians and other persons. These diplomas were framed and hung up in their offices. When the writer returned from Europe to Boston, in 1857, it was ascertained that there were nearly a hundred electropathist door-plates, or signs, of silver or on tin, to be seen throughout nearly all the streets of the city, on doors or windows of the offices of men or women electropathists. At that time there were nearly half as many of homœopathists, and as many more of hydropathists. Now, nearly all of which, of every sort, have disappeared. There are a few homœopathists left. Still, electricity, from suitable apparatus and skilfully applied, is a valued therapeutic in certain cases. And still, water, either hot or cold, is applied or injected in fevers, congestions, and inflammations as a valuable aid or remedy. And fine medicines,—as extracts or alkaloids, also in the form of coated pills or parvules, and powders, elegant in appearance and easily taken,—are now employed in the place of crude drugs by regular physicians.

Alexander H. Stephens, M.D., formerly president of and professor in the old College of Physicians and Surgeons of New York, my teacher, delivered an address before the State Medical Society and the Legislature at Albany, when he said: " We claim to be the exclusive depositors of sound medical knowledge, because we alone seek it at the only true and legitimate scource. Beginning with the epochs coeval with the latter historical times of the Old Testament, the science of medicine and the art of healing are the accumulated experience and the wisdom of ages. Everything connected with the cure of disease has been laboriously examined. The smallest artery, vein, nerve, and all other fibres have been minutely described, and changes induced by disease have been investigated by the aid of chemistry, electricity, and the microscope. In no other science has new descriptions advanced so far. All the facts relating to remedies and the nature and cure of diseases are made more and more available to the student by careful arrangement.

"The vocation of the practising physician is the spirit of Christianity in action. It consists not alone in healing the sick, in soothing the afflicted, and recalling the wandering intellect, but also in cherishing a love of peace and moderation amongst all men, and in promoting moral and intellectual improvement. The practice of the healing art is an occupation intrinsically dignified. It cannot be divested of this quality by the humble condition of the practitioner, or by the repulsive nature of any of his duties; still less by the lowly condition of his patient. In the most abject human being the true physician recognizes a fellow-man; in the most exalted, nothing more. The offspring of the highest and the lowest, in the first moments of their existence, come under his care, alike naked and helpless. The screen which in after life conceals many of their weaknesses and some of their virtues, ever open more or less to the medical observer, is for him removed by sickness and misfortune. Before the man of healing the trappings of greatness are laid aside and the cloak of deformity is dropped; beauty puts off her ornaments, and without a blush modesty raises her veil. And when, at last, man is about to take his departure into the abyss of eternity, he strips off all disguise and stands revealed in his primitive nakedness and helplessness. Surely those who hold such relations to society should be learned, honest, discreet, and wise ; trained by pure liberal studies and by illustrious examples to be ever true to science and humanity elevated by education and Christianity to rise above all that is low or sordid." There is no longer any more recognition of the real old-school doctor and his mysterious so-called remedies than there is of the *assumptions and assertions* of the alchemist of the sixteenth or seventeenth century. The fact is, homœopathy may well be recognized as the popular medical delusion of the day, and put on record as the *medical myth or mythology* of the nineteenth century.

It is generally known, or ought to be, that the great regular medical profession is in one complete fellowship in all this country and throughout all countries, and hence is completely *inclusive* of all true male and female physicians and remedies; but it will not accept nor in any way recognize any sort of sect, ism, or pathy. It now embraces, and has during the nineteenth century, all best medicines, means, and methods for preventing disease among the people, for relieving suffering, for healing the sick and wounded, for preserving life and limb and the reason of mind. All that is best and good found in the various ancient medical sects and schools, or is found in any more recent research, or taken from any quack or granny, is thankfully received and appropriated, because it belongs to the all-including regular medical profession. If there is any question or difference of opinion in regard to any medicine, dose, or method, or other important matter, we have clinics and medical journals, and societies for observation and for improvement, where fair and gentlemanly discussion can be had in the presence of facts and science.

At the last year's annual meeting of the Massachusetts Medical Society an address was given by Dr. H. C. Brown, who said: " We are fellows of a noble line of ancestors, medical men of the old school, aye, as old as when the first child was born in the Garden of Eden and our first parents ministered to the aches and ails of their children—men of the new school, yes, a broad school, old and new, which embraces every invention, discovery, and suggestion which can minister to the good of mankind, from whatever source it comes,—so long as it proves worthy and the best,—down to the latest moment of recorded time." Such true medical science is now conceded by all culture to be the highest and the noblest of the sciences.

What reason, then, it may be asked again, is there for the present existence of any such narrow, exclusive dogmatic pathy or ism? Whatever science and common sense and experience approves is more than welcome, and all else is rejected. Then why does homœopathy exist, since it is not recognized as a part of the regular medical profession ? Simply, the same as many other doubtful things exist, namely, because they are patronized without being thoroughly understood. It may be asked why quackery or Mormonism exists. Because there are classes of people who condone or patronize them, and others proselyte for them; and yet all good and thoughtful people lament their existence.

However, in the last dozen or twenty years, since the war, this sect called homœopathists has evidently been on the wane, as to unity or uniformity and *pure* homœopathic practice, as indeed it had some time before been losing largely in numbers and influence also throughout Great Britain, and before that in Paris, Berlin, Vienna, and all Europe. That is, where it began to flourish, had its head, and was best known, there it began first to decay and die. Such is true also of a certain class of animals; their tail will continue to wiggle and even make more flourish than natural for a long time after the head is cut off and the body is dead. This, then, must be a "symptom," and we may regard it as significant.

The vigilance of the homœopathists in proselyting is something wonderful. "Yes, homœopathy is in Congress, and will be in every department by and by.* The effort to get homœopathic physicians in the army and navy may fail, but it looks now as if the pressure would be very great—possibly successful. There is no use disguising the fact that the record of the surgical exploit in the case of President Garfield is having a depressing effect on the regular school. . . . Allopathy is under a cloud, and now is the time to push *our cause* at all points. It is to be hoped that every one who reads this will see the advantage to be gained, not only in Congress, but also in every community, by widely circulated petitions. Those who have not received blank copies of the petition should write to Dr. —— and get one and start it among their patients and candid people. It will give *the cause* a boom all around. The American love of fair play and equal rights will aid this cause wonderfully. Just the present status of the bill we are not able to ascertain. . Persistent united efforts will move Congress."

"The fair for the Garfield Monumental Hospital was not such a grand success as some hoped it would be. The fact that homœopathy is to be given no fair showing in it, and the other fact, of the movement to establish a homœopathic hospital in this central city, has had its influence. Such a hospital is needed . . . and should have help from Congress. Homœopathy is well represented here. Most of the physicians here are persons of strong individuality, able representative men. Most of them have widespread circles of admirers, and the lady members of our ranks here are both able and influential."

The need of such information as this little work is intended to give to general readers is now again made manifest. The Boston *Daily Advertiser* of August 6, 1884, has in a report of State House matters the following: "A scene of more than ordinary significance occurred at the office

* See *U. S. Homœopathic Medical Investigator*, January, 1883.

of the Secretary of State yesterday. . . . The establishment of a state homœopathic hospital is a direct recognition by the state of the homœopathic school of medicine, in the face of persistent and protracted opposition from the allopaths. After the homœopathic trustees had taken the oath of office four of them met in Room B. and chose a president and secretary." Thus we see also that the reporter speaks familiarly of the supposed two schools, homœopathy and allopathy, as if their existence were actually true; for such is the too general impression in certain portions of the community, and this accounts for such public recognition.

The real facts were, that some of the members of the Legislature were homœopathists, and that a leading professor in that school repeatedly appeared before that representative body and claimed that as a portion of the people are homœopathists therefore they should have control of some of the state hospitals, and that this professor's statements were entertained. Suppose the Roman Catholics, or the Methodists, or Baptists, should make the same plea for their exclusive control of some of the public schools, or the botanical doctors for exclusive state hospitals, would such propositions be entertained? Why not, if an exclusive sect of doctors are so favored, why not others and the various denominations of Christians? Such a mistake could occur only from the too prevalent impression that homœopathy is just as much a scientific theory and practice of medicine as is that of the great regular medical profession. As Massachusetts has lately become noted for having had a " very peculiar " governor, so now we are to be distinguished for having, at this day, a state homœopathic hospital for the insane! Is it not about time for us to expect applications for an electropathic hospital, and an allopathic or hydropathic state hospital, and perhaps a branch of Mormonism, within the commonwealth of Massachusetts?

We have another prominent example to the same effect. A few years ago there was a very wealthy and eccentric business man who was a strenuous advocate of homœopathy, but he knew more about fish and finance than he did about the learned professions, or *potency* in homœopathy, or the practice of medicine. He was connected with one of the most flourishing and excellent religious denominations, who, in their prosperity, determined to found an university in the centre of these Eastern States. This wealthy man was elected a trustee, together with other non-professional men, some of whom also happened to be homœopathists, or rather their wives were. So that in making up funds and arranging things for this new school the eccentric man proposed to endow the university, or the medical department of it, with a large sum, provided they adopted homœopathy, and had it taught to male and female students. This large sum of money was needed and it carried the day; so that that denomination and their excellent university, with their constellation of eminent professors of classical and modern literature, science, and philosophy, theology and law, have had foisted upon them this department of *sophistry!*

In times past some physician put upon record that he thought "the practice of medicine is an uncertain science." This is still often quoted at the present day, as if true as it was ages ago. And it is made an excuse for trying medicines made under the "certain law of homœopathy," which are said also to be so pure, exact, and reliable ! If these statements were only true there would be reason in it. But one might as well put to sea on a raft of straw because a Hahnemann says it is the safest way to cross the ocean. Suppose an old sea captain, in a moment of depressed mood, perhaps after he had lost his ship, should give as his opinion from experience "that navigation in the long run is an uncertain science," because on approaching the coast there often is thick heavy weather, snow storms, or fogs, or squalls, or heavy sea, a gale of wind, and that perhaps ahead, and who could tell just where he is ? Even if he could call a pilot (the specialist) who is perfectly familiar with the rocks and shoals about the entrance of that harbor, yet he is not always certain of getting safely in, for he cannot see his bearings. He might examine the reckoning and cast the lead to find soundings and the tide, and the sand or color of mud it brings up, and so put the helm to port or to starboard for bringing the ship into the channel, and yet he will sometimes fail. Besides, there are hulks adrift at sea as well as icebergs, and the ocean is dotted over with innumerable ships and steamers going very swiftly in all directions, in the darkest nights as well as days, and there are accidents liable to occur on board the ship herself, therefore "navigation is a very uncertain science." The fact is, thousands of people, less spleeny, trust themselves to it, not only from business necessity but also for pleasure, because for the most part it is not uncertain, but very certain. So, in all curable cases, is the regular practice of medicine by a competent and skilful physician.

The venerable Dr. John L. Atlee, of Philadelphia, in the course of his presidential address before the American Medical Association, held last spring at Cleveland, Ohio, said : "I have been a graduate in medicine sixty-three years. In my early day, previous to the establishment of medical societies throughout the country, and the organization of this great Association and the general adoption of the Code of Ethics, I saw many disastrous effects from the want of union and brotherly consideration and kindness. The medical men of that day were often in trouble and difficulties ; patients were sometimes transferred without ceremony, and so great was the jealousy existing among them no medical society could be formed for twenty years in my native city. Instead of being taken by the hand by the older physicians, every obstacle was thrown in my way,—consultations were refused, and the treatment of my patients unfavorably criticised. By the organization of medical societies, and adoption of the Code of Ethics, a wonderful change has been effected. We now feel it our duty and privilege to sustain our younger brethren, to treat them with courtesy and kind-

16

ness, to save them from their errors, and encourage them in all their good work. Had the adoption of the Code of Ethics no other result than this, it would have been a blessing to the medical profession. But it has accomplished more. *It has put a seal of condemnation on all ' isms,'* and developed an *esprit de corps* that has purified and enlarged the boundaries of our science, and greatly increased the usefulness and the social standing of the profession. One word more and I have done. I am completely satisfied, near the close of a long life, that in no other career can a man more fully accomplish his whole duty to God and to his fellow men; so that when life here is ended, it can be truly said of him,—and be it said with all reverence, —as it was said of Him whom we should imitate, *pertransivit benefaciendo,* "he went about doing good." On this line and to this end every true physician feels it to be his high privilege, as well as duty, to unswervingly purpose and work.

An art then, of all others the most noble, will be still cultivated by men of erudition and judgment, to the great benefit of mankind.

www.ingramcontent.com/pod-product-compliance
Lightning Source LLC
Chambersburg PA
CBHW030808020726
47499CB00006B/1821